Production and Maintenance Optimization Problems

Series Editor

Hisham Abou-Kandil

Production and Maintenance Optimization Problems

Logistic Constraints and Leasing Warranty Services

Nidhal Rezg
Zied Hajej
Valerio Boschian-Campaner

iSTE

WILEY

First published 2016 in Great Britain and the United States by ISTE Ltd and John Wiley & Sons, Inc.

ISTE Ltd
27-37 St George's Road
London SW19 4EU
UK

www.iste.co.uk

John Wiley & Sons, Inc.
111 River Street
Hoboken, NJ 07030
USA

www.wiley.com

Library of Congress Control Number: 2016949449

British Library Cataloguing-in-Publication Data
A CIP record for this book is available from the British Library
ISBN 978-1-78630-095-9

Contents

Introduction

I.1. Motivation and literature review

The improvement of industry involves the reduction of costs and maximization of customer satisfaction. Satisfying customer demands in a timely fashion has become difficult due to the random nature of such demands, a problem compounded by machine failures and low system availability. High system availability, minimal machine failure and customer satisfaction cannot be achieved without good management and a good knowledge of how to address problems while making decisions. These decisions are generally associated with three levels of hierarchical planning: strategic, tactical and operational planning.

The allocation of resources can become necessary over long periods of time, as purchase costs can become prohibitive. Subcontracting and leasing have become very important for many manufacturing enterprises because of the advantages that these solutions can bring. Such industrial solutions are becoming increasingly popular, for example subcontracting the workforce to perform certain tasks (maintenance, supervision, audit, etc.) or leasing workstations in order to produce the required quantities of products.

Several industrial constraints imposed on companies have led to the revision of integrated maintenance production strategies. Such strategies are adopted in order to develop and optimize new, integrated maintenance-based production strategies, taking into account certain industrial constraints, such as logistics, quality, warranties, and subcontracting. Through the development of such maintenance/production strategies produced under constraints, we can gain an overview of the maintenance strategies and production decisions required to balance industrial system availability, productivity and customer satisfaction.

I.2. Overview of the topic

This book explores several maintenance and production optimization problems, taking into account certain industrial constraints.

Chapter 1 covers an integrated production and maintenance optimization strategy for a forecasting production and maintenance problems. The production system is composed of a single machine M_1 subjected to random failure. In order to satisfy the random demand, under given service level, subcontracting assures the rest of the production through machine M_2 with transportation delay. An analytic study of the problem has been proposed using a sequential determination of the economical production plan for which an optimal preventive maintenance strategy has been calculated based on minimal repair. Firstly, an economic production plan of principal and subcontracting machines was obtained, which minimizes the total cost of production and inventory for the cases with and without returned products under service level and subcontracting transportation delay. Secondly, a joint maintenance strategy is determined according to the optimal production plan, under various constraints for production rates, transportation delay and returned production deadlines. Numerical results are presented to highlight the application of the developed approach and sensitivity analyses show the robustness of the model.

Chapter 2 presents a stochastic production, maintenance and delivery problem for a deteriorating manufacturing system. Under stochastic demand, in terms of service level, product return and delivery time, this book proposes a mathematical formulation based on quadratic modeling. Production and maintenance policies are developed in order to study the influence of delivery time on the planning of production, maintenance and delivery activities. Simulation results are presented to illustrate the exploitation of the proposed approach.

In Chapter 3 we develop a mathematical model based on the forecasting production/maintenance optimization problem, to study lease contracts with basic and extended warranties based on win-win relationship between the lessee and the lessor. The influence of production rates in equipment degradation and consequently on the total cost by each side during the finite leasing period is stated in order to determine a theoretical condition under which a compromise-pricing zone exists under different possibilities of maintenance policies.

Chapter 4 presents presents a control policy of a manufacturing system under cost, availability and quality constraints. The production system consists of a two machines and two buffers and produces conforming and non-conforming products. A preventive maintenance strategy is developed in order to determine the instants at which preventive maintenance has to be performed on each machine, and both buffer inventory levels. A simulation, experimental design and multi-criteria analysis are presented to prove the adopted approach.

1

Forecasting and Maintenance under Subcontracting Constraint with Delay in Transportation

This chapter presents a forecasting problem relating to production and maintenance optimization to meet random demand with a single machine M_1 on a finite horizon. The function rate of M_1 depends on the production rate for each period within the forecasting horizon. In order to satisfy customer demand, subcontracting assures the remaining production through machine M_2 with a delay in transportation. An analytical formulation of the problem is proposed using sequential computation of the optimal production plan, for which an optimal preventive maintenance policy has been calculated based on minimal repair.

First, we find, the optimal production plans of the principal (M_1) and subcontracting (M_2) machines. Such plans minimize the total production and inventory cost for situations with and without returned products at an agreed service level and with a delay in subcontracting transportation.

Second, we determine a joint effective maintenance policy with the optimal production plan, which integrates the various constraints for production rates, transportation delay and returned production deadline.

Numerical results are presented to highlight the application of the approach we develop and sensitivity analysis shows the robustness of the model.

1.1. Introduction

Industry improvement requires a reduction in costs and maximization of customer satisfaction. These two goals can be achieved with good management and decision-making. The importance of subcontracting has grown both from economic and production points of view. The new manufacturing paradigm, which emphasizes outsourcing, cooperation, networking and agility, is regularly discussed at a general level but very little empirical research has been done on these issues.

Amesse *et al.* [AME 01] introduced the importance of subcontracting as an industrial strategy across all domains. Subcontracting requires collaboration, logic, coordination and management between the manufacturing companies in order to meet customer requires in terms of quantity and delay [AND 99, BER 01].

Recently, more work relating to production and maintenance coupling has been published that integrates new constraints corresponding to the concept of subcontracting. There are a number of different works that deal with subcontracting under constraints, for example [DEL 07] and [DAH 10]. Dellagi *et al.* [DEL 07] have contributed to the development of integrated maintenance policies while coupling maintenance and production under the constraint of subcontracting. In an industrial model, they assumed that production consisted of only one machine and, in order to satisfy customer demand, it was necessary to collaborate with another subcontracting machine. Dahane *et al.* [DAH 10] aims to determine maintenance policies that consider the concept of subcontracting, but concerning the provider of a subcontracting service. The optimal time for maintenance and the optimal stock level, considering the relationship between production and maintenance, is determined. The demand, in several works that take the subcontracting approach, is assumed to be constant and known across an infinite horizon. This type of problem is more difficult in the case of random demand over a finite horizon. In this situation, variations in production rates are necessary to meet demand.

Regarding a production/inventory problem without maintenance, Holt *et al.* [HOL 60] proposed a model defining a quadratic cost minimization program that approximates the cost functions for hiring and laying off labor, overtime, inventory and product shortage through the use of suitable quadratic functions. As a result, and considering some constraints, this model provides an optimal smoothing solution for aggregating inventory, production and the workforce. In this context, Silva and Cezarino [SIL 04] have analyzed a production–planning optimization problem that uses both imperfect information from decision inventory variables and computes the expected cost.

Several works have dealt with the interdependent relationship between production and maintenance planning. There are different attempts to study the problem of conflict in management decisions and the necessity of combining objectives in order to enhance the global benefits of industry, and mainly to minimize global costs, in the literature. Research has been carried out to analyze the problem of joint production and maintenance optimization. In this context, Aghazzaf *et al.* [AGH 08] have developed models dealing with integrated maintenance based on aggregated production planning, where decision variables related to preventive and corrective maintenance are used. Recently, Hajej *et al.* [HAJ 11] have dealt with combined production and maintenance plans for a manufacturing system satisfying random demand over a finite horizon. In their model, they consider the influence of production on the degradation of a machine, and consequently consider maintenance planning.

In our study, we build on models presented in Hajej *et al.* [HAJ 09] and Ayed *et al.* [AYE 12] where the given manufacturing systems cannot ensure the total demand over the given time horizon and subcontracting is called for.

Ayed *et al.* [AYE 12] dealt with a randomly failing manufacturing system M_1 which has to satisfy random demand across a finite horizon at a required service level. To help meet demand, subcontracting through another production system M_2 is used. M_1 operates with a variable production rate and its failure rate depends on both time and production rate.

Hajej *et al.* and Ayed *et al.* [AYE 12 HAJ 09, HAJ 11, HAJ 12] have, however, ignored several significant characteristics and terms of manufacturing systems, such as transportation, terms of delay, quantity and subcontracting transport, in their work. Many pieces of research analyze transportation delays, such as the delay in delivery between a manufacturing plant and the warehouse that has purchased the manufactured goods, and the impact of such delays on the manufacturing system. For example Richard and Chen [RIC 05], which considered a multi-agent architecture of supply chain integration, proposed heuristics and programming models in order to devise demand-driven supply chains via two types of bidding approaches: customization and webbing. Recently, based on the works by Hajej *et al.* [HA 11, HAJ 12], Turki *et al.* [TUR 12] studied a simple manufacturing model composed of one machine with a transport delay between production (at the manufacturing plant) and receipt by the customer (at the warehouse) by treating the impact of delivery time and withdrawal on production/maintenance planning and quantity transported per time period in order to satisfy a random demand.

Motivated by the work in Turki *et al.* [TUR 12], we treat the aspect of transportation in another more complex and realistic industrial system composed of two machines (a principal and a subcontractor machine), by integrating a subcontractor with its related characteristics, such as transportation delay. This study has novelty and originality in the development of a production and maintenance optimization plan to address this type of problem. It shows that a subcontractor machine can be used to help guarantee the desired service level by distributing production so that the principal machine is not used at its maximum rate, since its degradation rate is correlated with production level.

The primary objective of this chapter is to determine economical production planning over the finite horizon based on forecasting demand, taking into account the transportation delay relating to subcontracting. The impact of transportation delay on optimal production planning will be studied thereafter. Our secondary objective is to establish economical production plans for the principal and subcontractor machines, taking into account the influence of products returned to the production system. The last objective is to determine a joint effective maintenance policy using the optimal production plan, which integrates the various constraints for production rates, transportation delay and the deadline for product return.

This remainder of this chapter is organized as follows:

– section 1.2 states the problem in the case where there are no returned products. A general stochastic production/inventory model is formulated. The policy and analytical expression of production/inventory are developed considering the influence of subcontracting transportation delay on the production plans of principal and subcontracting machines;

– section 1.3 deals with the case where products are returned. It uses the initial system and shows the influence of the right to withdraw from the production system;

– section 1.4 presents and develops the policy and the analytical expression of maintenance, considering the influence of subcontracting transportation delay and the return of products on the optimal maintenance strategy;

– the conclusion is given in section 1.5.

1.2. Production without retuned products

1.2.1. *Statement of the problem*

In this work, optimal production planning based on forecasting the problem of demand is formulated. We consider a manufacturing system problem in which one part-type is produced through a single operation in order to satisfy random demand over a finite horizon, H. We assume that the fluctuation in demand follows a normal distribution with the mean and variance given by $\hat{d}$ and σ_d, respectively.

In order to satisfy this random demand with a given inventory service level α, and to avoid shortage due to manufacturing system unavailability, the enterprise has to build a stock buffer. The need for additional stock to be used as a buffer leads to the need to call upon another production enterprise, called the subcontractor. The transportation delay related to the subcontractor, denoted by τ, is therefore considered between the subcontractor machine and the stockpile, S. The products leaving the subcontractor are transported with a transportation delay τ before they arrive at the principal stockpile.

The use of a subcontractor machine means that the main machine does not have to work at maximum capacity, which reduces the total production, inventory and maintenance costs. The subcontractor machine is used to reduce pressure on the principal machine and therefore reduce the number of failures and the maintenance cost.

Maintenance of the subcontractor's manufacturing system, M_2, is outside our control. The only information about its maintenance is the availability rate, β_2. This assumption is realistic, since each piece of equipment has its own failure rate and therefore its availability can be calculated using the theory of reliability. In practice, availability is an indicator that companies are always trying to improve upon. Machine M_2 is also characterized by its maximal production rate U_2^{max} and its unit production cost C_{pr2}, with $C_{pr2} > C_{pr1}$. The industrial problem is illustrated in Figure 1.1.

Our objective is to establish an economical production plan that satisfies random demand, according to demand forecasting, and takes into account the subcontractor transportation delay. The aim is to minimize the sum of the production and inventory costs.

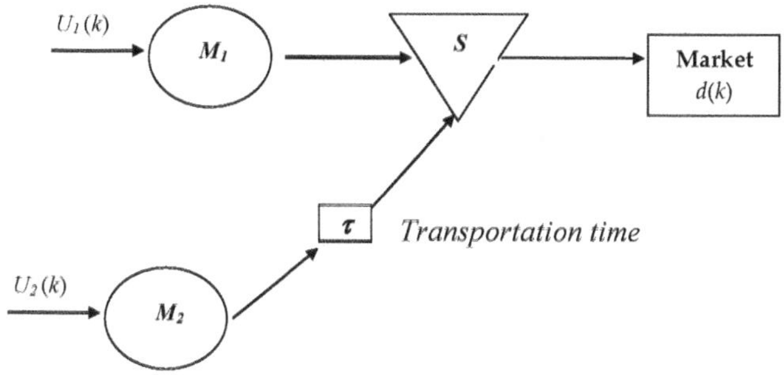

Figure 1.1. *Initial production problem without any returned products*

1.2.2. *Notation*

The following parameters are used in the mathematical formulation of the model:

τ	transportation delay
τ'	deadline for product return
Δt	length of a production period
H	number of production periods in the planning horizon
$H.\Delta t$	length of the finite planning horizon
$U_{i,k}$	production rate by machine M_i, $i\in\{1,2\}$ during period k $(k = 0, 1,\ldots, H-1)$
$\hat{d}(k)$	average demand during period k $(k = 0, 1,\ldots, H)$
$V_{d(k)}$	variation in demand during period k $(k = 0, 1,\ldots, H)$
S_k	inventory level of S at the end of period k $(k = 0, 1,\ldots, H)$
$\hat{S}_k$	average inventory level of S during period k $(k = 0, 1,\ldots, H)$
C_{pr1}	unit production cost of machine M_1

C_{pr2}	unit production cost charged by the subcontractor for machine M_2
Cs	holding cost of one product unit during one period
M_1	machine 1
M_2	machine 2
mu	monetary unit
U_1^{max}	maximum production rate of M_1
U_1^{min}	minimum production rate of M_1
U_2^{max}	maximum production rate of M_2
U_2^{min}	minimum production rate of M_2
α	probability index related to customer satisfaction and expression of the service level
β_2	M_2 availability rate
δ	percentage of products returned
S_0	initial inventory
ξ_M	total maintenance cost
C_{cm}	corrective maintenance cost
C_{pm}	preventive maintenance cost

1.2.3. *Optimization of production policy*

The principal idea is to minimize the expected production and inventory costs over a finite time horizon [0, H]. It is supposed that the horizon is divided equally into H periods [HAJ 11]. Demand is satisfied at the end of each period. The problem can be formulated as a linear–stochastic optimal control problem under the constraint of a stock level threshold, with the production rates corresponding to each period as the decision variables.

In the stochastic problem, $f_k(.)$ denotes functions that represent the production, and inventory costs, and $E\{\}$ denotes the mathematical average value operator. Referring to Hajej *et al.* [HAJ 11], we formulate the problem as follows:

$$\underset{U(k)}{Min}\left(E\left\{ \sum_{k=0}^{H-1} f_k\left(S_k, U_k\right) + f_H\left(S_H\right) \right\} \right) \qquad [1.1]$$

subject to the level of inventory S at period $(k+1)$. This is determined by calculating:

– the inventory of S during period k;

– the production rate of the principal machine M_1 during period (k);

– the production rate of the subcontractor machine M_2 during period $(k-\tau)$;

– the demand rate during period k.

Consequently, the inventory balance equation for each time period is formulated in this way:

$$S_{k+1} = S_k + U_{1,k} + \beta_2 \cdot U_{2,k-\tau} - d_k \text{ with } k \in \{0,1,...,H-1\} \qquad [1.2]$$

with τ being the transportation delay introduced by the subcontractor.

The quantity of products produced by the subcontractor that arrives at the principal stockpile S during period k is the quantity of products that has left the subcontractor during the period $k - \tau$. The transportation delay τ is linked to the length of the production period, i.e. the products arrive at the principal stockpile after τ production periods, so we can write this as follows:

$\tau.\Delta t$ with $\tau \in \{1,2,.....\}$

where Δt is the length of the production period.

To determine in which period the sub-contractor will begin delivery, we therefore need to calculate $k \in \{1,2,.....\} - \tau$.

The service level for each period is expressed by the following constraint.

$$\text{Prob}\left[S_{k+1} \geq 0\right] \geq \alpha \text{ with } k \in \{0,1,\ldots,H-1\} \quad [1.3]$$

The following constraint defines the upper and lower bounds of the production level during each period *k*:

$$0 \leq U_k \leq U_1^{max} + U_2^{max} \text{ with } k \in \{0,1,\ldots,H-1\} \quad [1.4]$$

1.2.4. *Total production and inventory cost*

In this section, we formulate a constrained stochastic problem meeting a required service level, including subcontractor transportation delay and random demand, by using the Holt–Modigliani–Muth–Simonmodel.

1.2.4.1. *The Holt–Modigliani–Muth–Simon model*

The Modigliani–Muth–Simon model is considered one of the first models to deal with the certainty–equivalence principle for dynamic linear quadratic problems [BER 95]. It is usually applied as a benchmarking tool in order to compare different production planning approaches and to provide managers and decision makers with perspectives on and ideas about how to manage a firm's material resources [SIN 96]. Some other works, such as Hax *et al.* [HAX 84], have proven that the Modigliani–Muth–Simon model is useful for evaluating the production process. So, for example, the quadratic inventory cost describes and takes into account both negative (rupture and backorders) and positive (overstocking) inventory status.

Inspired by the Modigliani–Muth–Simon model, we got the idea of moving the emphasis to the machines instead of the workers, production rate and inventory levels in order to plan optimal production. In our work we also make some changes to the model, keeping its linear quadratic form. Furthermore, we take into account some constraints on the decision variables in order to make our approach more realistic and to ensure its applicability to real industrial cases.

1.2.4.2. *Quadratic total cost*

In our problem, we reformed the Modigliani–Muth–Simon model to determine an inventory and production policy that respected the principle characteristic of an Modigliani–Muth–Simon model: the use of a quadratic cost function. This function allows both an excess of and shortage in inventory levels to be penalized.

The expected production and inventory cost for period k is given by:

$$f_k\left(U_{1,k},U_{2,k},S_k\right)=C_s E\left\{S_k^2\right\}+C_{pr1}U_{1,k}^2+C_{pr2}\beta_2 U_{2,k}^2 \quad [1.5]$$

The total expected cost of production and inventory over the finite horizon $H.\Delta t$ can then be expressed as follows:

$$F(u)=\sum_{k=0}^{H} f_k\left(U_{1,k},U_{2,k},S_k\right)=C_s E\left\{S_H^2\right\}+\sum_{k=0}^{H-1}\left[C_s E\left\{S_k^2\right\}+C_{pr1}U_{1,k}^2+C_{pr2}\beta_2 U_{2,k}^2\right]$$

with $k\in\{0,1,\ldots,H-1\}$ [1.6]

The decision to square two variables – storage and production – is justified as it reflects the variation in stock storage and shortage. This is an approximation used by economists.

1.2.5. *Analytical study*

In this section, we show the transformation of a stochastic problem using the analytical study of policy and establish the deterministic equivalent problem.

1.2.5.1. *Production and inventory costs*

This approach consists of transforming the stochastic problem into a deterministic equivalent by maintaining the principal properties of the original problem.

Firstly, we propose the following notation for the mean variables:

$$E\{S_k\} = \hat{S}_k \,,\; E\{U_{1,k}\} = U_{1,k} \,,\; E\{U_{2,k}\} = U_{2,k}$$

U_k being deterministic for each interval Δt, since it does not depend on the random variables d_k and $\hat{S}_k$.

Thus $\hat{U}_k = U_k$:

and $Var_{U_k} = 0$:

with $U_k = U_{1,k} + \beta \cdot U_{2,k-\tau}$.

The total production and inventory cost are as follows in Lemma 1.1.

LEMMA 1.1.–

$$F(u) = C_s \hat{S}_H^2 + \sum_{k=0}^{H-1} \left[C_s \hat{S}_k^2 + C_{pr1} U_{1,k}^2 + C_{pr2} \beta_2 U_{2,k}^2 \right] + C_s \cdot \sigma_d{}^2 \cdot \frac{H(H+1)}{2} \qquad [1.7]$$

PROOF (see Appendix 1).–

The inventory balance equation is as follows:

$$\hat{S}_{k+1} = \hat{S}_k + U_{1,k} + \beta_2 \cdot U_{2,k-\tau} - \hat{d}_k \text{ with } k \in \{0,1,\ldots,H-1\}$$

1.2.5.2. *Service level constraint*

To continue transforming the stochastic problem into an equivalent deterministic one, we consider service level constraint in a deterministic form by specifying a minimum cumulative production quantity depending on the service level requirements through the following lemma.

LEMMA 1.2.–

We recall that α defines the service level constraint. This constraint is expressed as follows:

$$\mathrm{Prob}\left[S_{k+1} \geq 0\right] \geq \alpha \text{ with } 0 \leq U_k \leq U_1^{\max} + U_2^{\max}$$

Then, for $k = 0,1,..,H-1$, we have:

$$U_k \geq U_\alpha(S_k, \alpha) \text{ with } U_k = U_{1,k} + \beta_2 \cdot U_{2,k-\tau} \qquad [1.8]$$

where $U_\alpha()$ represents the minimum cumulative production quantity expressed as follows:

$$U_\alpha(S_k, \alpha) = V_{d_k} \varphi_{d_k}^{-1}(\alpha) + \hat{d}_k - S_k \text{ ; } k = 0,1,\ldots,H-1,$$

where:

– V_{d_k} is variation in demand d during period k;

– φ_{d_k} is the cumulative Gaussian distribution function with mean $\hat{d}_k$ and finite variance $V_{d_k} \geq 0$;

– $\varphi_{d_k}^{-1}$ is the inverse distribution function.

PROOF.–

$$\mathrm{Prob}\left[S_{k+1} \geq 0\right] \geq \alpha \text{ with } 0 \leq U_k \leq U_1^{\max} + U_2^{\max}$$

$$U_k = U_{1,k} + \beta_2 \cdot U_{2,k-\tau}$$

$$\Rightarrow \mathrm{Prob}\left[S_k + U_{1,k} + \beta_2 \cdot U_{2,k-\tau} - d_k \geq 0\right] \geq \alpha$$

$$\Rightarrow \text{Prob}\left[S_k + U_{1,k} + \beta_2 \cdot U_{2,k-\tau} \geq d_k \right] \geq \alpha$$

$$\Rightarrow \text{Prob}\left[S_k + U_{1,k} + \beta_2 \cdot U_{2,k-\tau} - \hat{d}_k \geq d_k - \hat{d}_k \right] \geq \alpha$$

$$\Rightarrow \text{Prob}\left[\frac{S_k + U_{1,k} + \beta_2 \cdot U_{2,k-\tau} - \hat{d}_k}{V_{d_k}} \geq \frac{d_k - \hat{d}_k}{V_{d_k}} \right] \geq \alpha$$

This equation is in the form of Prob $[Y \geq X]a$, with $X = \frac{d_k - \hat{d}_k}{V_{d_k}}$ being a Gaussian random variable representing demand, d_k, and where φ_{d_k} is a cumulative Gaussian distribution function of the following form:

$$\varphi_{d_k}\left(\frac{S_k + U_{1,k} + \beta_2 \cdot U_{2,k-\tau} - \hat{d}_k}{V_{d_k}} \right) \geq \alpha \qquad [1.9]$$

Since $\lim\limits_{d_k \to -\infty} \varphi_{d_k} = 0$ and $\lim\limits_{d_k \to +\infty} \varphi_{d_k} = 1$, function φ_{d_k} is strictly increasing, and we note that it is indefinitely differentiable. That is why we conclude that φ_{d_k} is invertible.

Thus, equation [1.9] becomes:

$$\frac{S_k + U_{1,k} + \beta_2 \cdot U_{2,k-\tau} - \hat{d}_k}{V_{d_k}} \geq \varphi_{d_k}^{-1}(\alpha) \qquad [1.10]$$

$$\Rightarrow S_k + U_{1,k} + \beta_2 \cdot U_{2,k-\tau} - \hat{d}_k \geq \varphi_{d_k}^{-1}(\alpha) \cdot V_{d_k}$$

$$\Rightarrow U_{1,k} + \beta_2 \cdot U_{2,k-\tau} \geq \varphi_{d_k}^{-1}(\alpha) V_{d_k} - S_k + \hat{d}_k$$

It can consequently be concluded that:

$$U_\alpha\left(S_k, \alpha\right) = V_{d_k}\, \varphi_{d_k}^{-1}(\alpha) + \hat{d}_k - S_k \quad with\ k = 0,1,\ldots,H-1.$$

Using Lemma 1.1 and Lemma 1.2, we resume the equivalent deterministic model as follows:

$$F(u)=C_s\hat{S}_H^2+\sum_{k=0}^{H-1}\left[C_s\cdot\hat{S}_k^2+C_{pr1}\cdot U_{1,k}^2+C_{pr2}\cdot\beta_2\cdot U_{2,k}^2\right]+C_s(\sigma_d)^2\frac{H(H+1)}{2}$$

under the following constraints:

- $\hat{S}_{k+1}=\hat{S}_k+U_{1,k}+\beta_2\cdot U_{2,k-\tau}-d_k$ with $k=\{0,1,\ldots,H-1\}$;
- $U_{1,k}+\beta_2\cdot U_{2,k-\tau}\geq\varphi_{d_k}^{-1}(\theta)V_{d_k}-S_k+\hat{d}_k$ with $k=\{0,1,\ldots,H-1\}$;
- $0\leq U_k\leq U_1^{\max}+U_2^{\max}$ with $k=\{0,1,\ldots,H-1\}$;
- $0\leq U_{1,k}\leq U_1^{\max}$;
- $0\leq U_{2,k}\leq U_2^{\max}$;
- $U_k=U_{1,k}+\beta_2\cdot U_{2,k-\tau}$.

1.2.6. *Numerical example*

In this section, a number of numerical examples are presented in order to illustrate the use of the mathematical model developed in the previous sections. We assume that the finite planning horizon $H.\Delta t$ is equal to 24 production periods with period length $\Delta t=1$month.

Subcontractor machine data	
availability rate	$\beta_2=0.93$
Unit production cost	$C_{pr2}=10$ *mu*
Maximal production rate	$U_2^{Max}=130$ unit
Principal machine data	
Unit production cost	$C_{pr1}=3$ mu/unit
Maximal production rate	$U_1^{Max}=130$ unit
Inventory data	
Inventory holding cost	Cs = 5 mu/Δt
Initial inventory level	S(0) = 200
Service level	$\alpha=0.9$

Table 1.1. *Numerical data*

Example 1

For the first example, the random demand is characterized by a standard deviation that equals $\sigma = \sqrt{V_{d_k}} = 5$ and the average demand for each period is presented in Table 1.2 below. This example treats the case where the manufacturing problem does not consider the transportation delay of subcontracting (*τ = 0)*.

d_0 = 150	d_1 = 170	d_2 = 150	d_3 = 150	d_4 = 150	d_5 = 140	d_6 = 160	d_7 = 140
d_8 = 160	d_9 = 130	d_{10} = 150	d_{11} = 140	d_{12} = 150	d_{13} = 120	d_{14} = 150	d_{15} = 130
d_{16} = 150	d_{17} = 110	d_{18} = 160	d_{19} = 130	d_{20} = 150	d_{21} = 120	d_{22} = 140	d_{23} = 160

Table 1.2. *Mean demand*

In order to realize this optimization, we use the Numerical Algorithms for Constrained Global Optimization with the MATHEMATICA software. The economical production plans for the principal and subcontractor machines are presented respectively in Tables 1.3 and 1.4 with a minimal total cost equals to 2.9736×10^6 mu.

$U_{1,0}$ = 107	$U_{1,1}$ = 55	$U_{1,2}$ = 120	$U_{1,3}$ = 59	$U_{1,4}$ = 95	$U_{1,5}$ = 124	$U_{1,6}$ = 107	$U_{1,7}$ = 80
$U_{1,8}$ = 114	$U_{1,9}$ = 81	$U_{1,10}$ = 33	$U_{1,11}$ = 58	$U_{1,12}$ = 62	$U_{1,13}$ = 94	$U_{1,14}$ = 13	$U_{1,15}$ = 46
$U_{1,16}$ = 127	$U_{1,17}$ = 82	$U_{1,18}$ = 115	$U_{1,19}$ = 22	$U_{1,20}$ = 53	$U_{1,21}$ = 6	$U_{1,22}$ = 71	$U_{1,23}$ = 113

Table 1.3. *Principal machine: $U^*_{1,K}$*

$U_{2,0}$ = 21	$U_{2,1}$ = 42	$U_{2,2}$ = 31	$U_{2,3}$ = 47	$U_{2,4}$ = 86	$U_{2,5}$ = 17	$U_{2,6}$ = 86	$U_{2,7}$ = 28
$U_{2,8}$ = 120	$U_{2,9}$ = 19	$U_{2,10}$ = 116	$U_{2,11}$ = 57	$U_{2,12}$ = 92	$U_{2,13}$ = 60	$U_{2,14}$ = 119	$U_{2,15}$ = 84
$U_{2,16}$ = 128	$U_{2,17}$ = 119	$U_{2,18}$ = 80	$U_{2,19}$ = 49	$U_{2,20}$ = 62	$U_{2,21}$ = 45	$U_{2,22}$ = 74	$U_{2,23}$ = 62

Table 1.4. *Subcontractor machine: $U^*_{2,k}$*

Example 2

This second example deals with the impact of the transportation delay of subcontracting on the production policy such as the economical production plans for the principal and subcontractor machines by varying the value of τ. We take the same average demand presented in Table 1.2. Tables 1.4 and 1.5 presented the economical production plans for the principal and subcontractor machines in the case where the transportation delay of subcontracting $\tau = 2.\Delta t$ with a minimal total cost equals 3.65355×10^6 um. In a more general way, Figures 1.2 and 1.3 show the variability of subcontracting transportation delay. In this case, as transportation delay τ increases, the production rates of the principal machine and subcontractor machines are increased in order to satisfy the given service level. Similarly, from Figure 1.4, we can note that if the transportation delay of subcontracting increases, then the total production/inventory cost increases.

$U_{1,0}$ = 113	$U_{1,1}$ = 63	$U_{1,2}$ = 120	$U_{1,3}$ = 110	$U_{1,4}$ = 130	$U_{1,5}$ = 42	$U_{1,6}$ = 60	$U_{1,7}$ = 70
$U_{1,8}$ = 120	$U_{1,9}$ = 116	$U_{1,10}$ = 68	$U_{1,11}$ = 103	$U_{1,12}$ = 73	$U_{1,13}$ = 99	$U_{1,14}$ = 62	$U_{1,15}$ = 13
$U_{1,16}$ = 120	$U_{1,17}$ = 90	$U_{1,18}$ = 121	$U_{1,19}$ = 32	$U_{1,20}$ = 130	$U_{1,21}$ = 70	$U_{1,22}$ = 130	$U_{1,23}$ = 98

Table 1.5. *Principal machine: $U^*_{1,k}$*

$U_{2,0}$ = 95	$U_{2,1}$ = 104	$U_{2,2}$ = 44	$U_{2,3}$ = 97	$U_{2,4}$ = 84	$U_{2,5}$ = 108	$U_{2,6}$ = 33	$U_{2,7}$ = 78
$U_{2,8}$ = 55	$U_{2,9}$ = 43	$U_{2,10}$ = 72	$U_{2,11}$ = 95	$U_{2,12}$ = 81	$U_{2,13}$ = 130	$U_{2,14}$ = 77	$U_{2,15}$ = 21
$U_{2,16}$ = 10	$U_{2,17}$ = 54	$U_{2,18}$ = 52	$U_{2,19}$ = 42	$U_{2,20}$ = 0	$U_{2,21}$ = 69	$U_{2,22}$ = -	$U_{2,23}$ = -

Table 1.6. *Subcontractor machine: $U^*_{2,k}$*

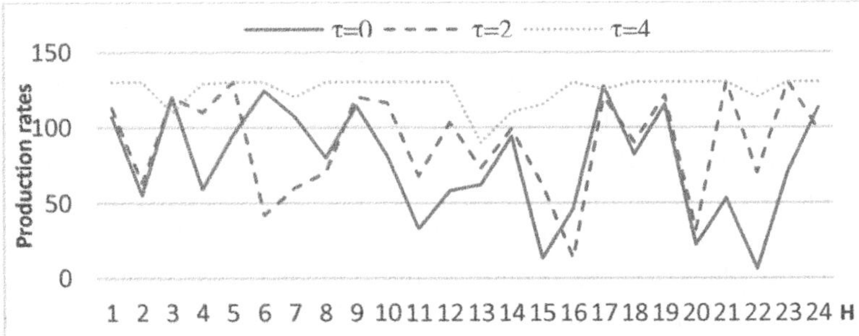

Figure 1.2. *Production rate variation as a function of τ for the principal machine*

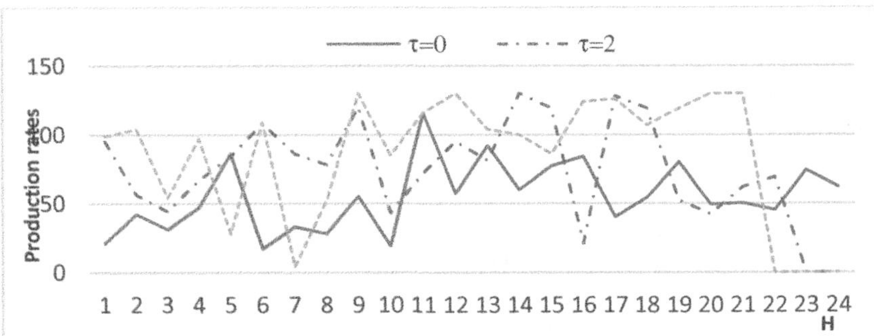

Figure 1.3. *Production rate variation as a function of τ for subcontractor machine*

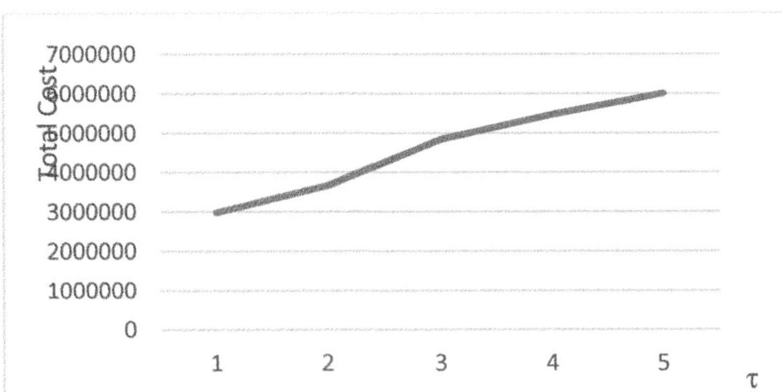

Figure 1.4. *Total cost variation as a function of τ*

Example 3

In the same context, we treat the influence of the service level on the production policy such as the optimal production plans for main and subcontractor machines as well as the minimal total cost. In this case, we use the same average demand of the first example and we take the case where the transportation delay of subcontracting is $\tau = 2$. From Figure 1.5 and Table 1.7, we interpreted that the higher value of production rates for the principal and subcontractor machines as well as the higher value of the total cost corresponds to the higher service level. This can be explained by the fact that, when the service level increases and in order to satisfy this higher service level, the main and subcontractor machines are required to produce more.

α	Total production and inventory cost
0.7	2.86485×10^6
0.9	3.65355×10^6
0.97	4.87342×10^6

Table 1.7. *Total cost variation as a function of service level α with $\tau = 2$*

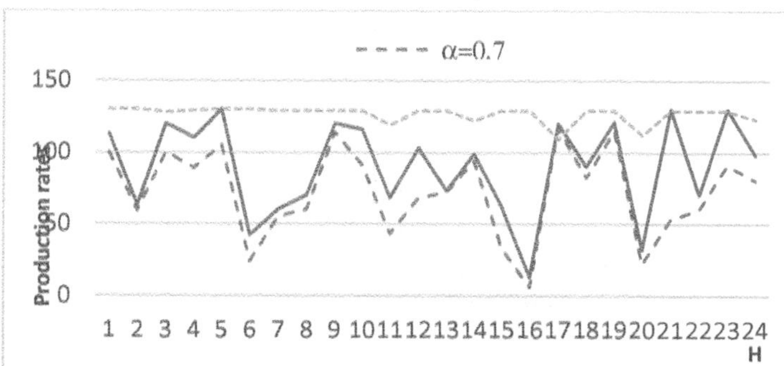

Figure 1.5. *Production rate variation as a function of service level for principal machine with $\tau = 2$*

Example 4

In this example, we are interested in finding the optimal production plans of principal and subcontracting machines for each value of standard deviation σ in order to study the impact of the standard deviation variation on the production policy. Using Table 1.8 and Figure 1.6, we can see that the higher value of the total production/inventory cost corresponds to the higher value of the demand variance variability. This can be clarified by the reality that, as variance of demand increases, the production rates of principal and subcontractor machines increase, the stock level augments and, consequently, the total production and inventory cost increases.

σ	Total production and inventory cost
1.5	3.36385×10^6
2.23	3.65355×10^6
5	5.34356×10^6

Table 1.8. *Total cost variation as a function of standard deviation σ with $\tau = 2$*

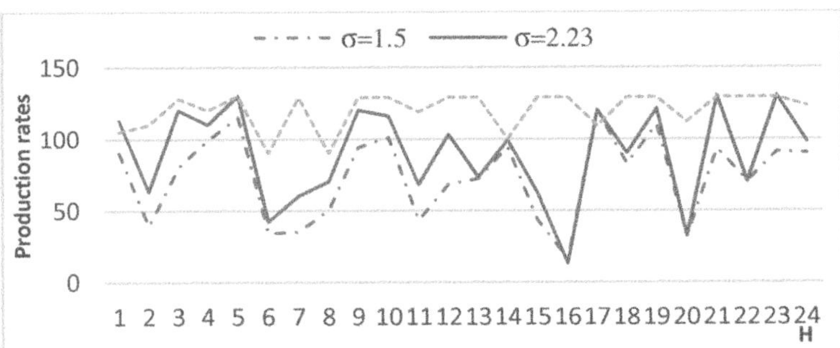

Figure 1.6. *Production rate variation as a function of standard deviation σ with $\tau = 2$*

1.3. Production with retuned products

1.3.1. *Statement of the problem*

In this section, we extend the initial system to include the case of returned products. In order to satisfy random demand with an inventory service level of α, the enterprise calls upon another subcontractor. At the same time, the production system takes into account products returned by customers. These products are still new and in stock. In the field of manufacturing, the return of products to stock is called the "right of withdrawal". This right gives the customer a specific deadline τ' by which to return products. Our objective is to establish economical production plans for the principal and subcontractor machines, taking into account the influence of the right of withdrawal in the production system.

The problem we will be modeling is illustrated in Figure 1.7.

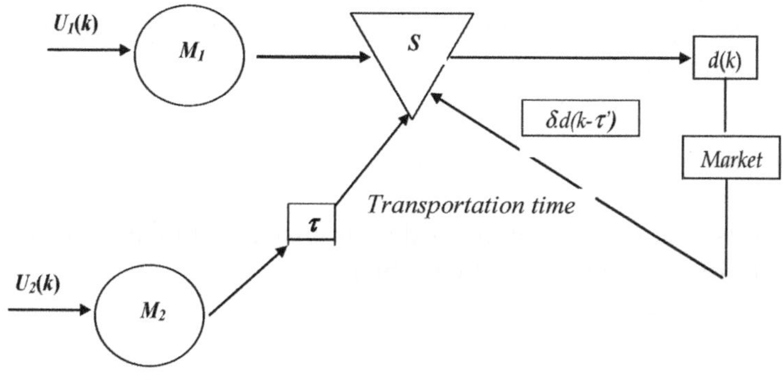

Figure 1.7. *Production problem with returned products*

1.3.2. Optimization of the production policy

The following stochastic problem provides optimal production plans for the planning horizon:

$$\underset{(U_1,U_2)}{Min}\left(\sum_{k=0}^{H} f_k\left(U_{1,k},U_{2,k},S_k\right)=C_s\cdot E\left\{S_H^2\right\}+\sum_{k=0}^{H-1}\left[C_s\cdot E\left\{S_k^2\right\}+C_{pr1}\cdot U_{1,k}^2+C_{pr2}\cdot\beta_2\cdot U_{2,k}^2\right]\right)$$

subject to:

$$S_{k+1}=S_k+U_{1,k}+\beta_2\cdot U_{2,k-\tau}+r(k)-d_k\ ,\ k\in\{0,1,\dots,H-1\} \quad [1.11]$$

$$r(k)=\delta.d_{k-\tau'},\ k\in\{0,1,\dots,H-1\} \quad [1.12]$$

$$\text{Prob}\left[S_{k+1} \geq 0\right] \geq \alpha \ , \ k \in \{0,1,\ldots,H-1\} \qquad [1.13]$$

$$0 \leq U_k \leq U_1^{\max} + U_2^{\max} \ , \ k \in \{0,1,\ldots,H-1\} \qquad [1.14]$$

As in section 1.2, constraint [1.11] defines the inventory balance equation for each time period. Relation [1.12] defines the quantity of products returned by the customer; this quantity is part of the demand returned by the customer after the specific deadline, τ'. Constraint [1.13] imposes a service level requirement on each period as well as a lower bound on the inventory variables so as to prevent stock-outs. The last constraint defines an upper bound on the production level during each period, *k*. We therefore cannot exceed a given maximum production rate.

1.3.3. *Analytical formulation*

In this section, we devise a deterministic formulation to make it easier to resolve our stochastic problem.

1.3.3.1. *Production and holding costs*

We can simplify the expected value of the production/inventory costs as follows in Lemma 1.3.

LEMMA 1.3.–

$$F(U_1,U_2) = C_s \cdot \hat{S}_H^2 + \sum_{k=0}^{H-1}\left[C_s \cdot \hat{S}_k^2 + C_{pr1} \cdot U_{1,k}^2 + C_{pr2} \cdot \beta_2 \cdot U_{2,k-\tau}^2\right] + C_s \cdot \sigma_d^2 \cdot \frac{(H+1)}{2} \cdot \left[H + \delta^2 \cdot (H-2)\right] \qquad [1.15]$$

1.3.3.2. *The inventory balance equation*

If we have $d_k = \hat{d}_k$, the state balance equation for stock (equation [1.11]) can be converted to:

$$\hat{S}_{k+1} = \hat{S}_k + U_{1,k} + \beta_2 \cdot U_{2,k-\tau} + \hat{r}_k - \hat{d}_k \ \text{with } k \in \{0,1,\ldots,H-1\} \qquad [1.16]$$

$$\hat{r}_k = \delta . \hat{d}_{k-\tau'} \text{, with } k \in \{0,1,\ldots,H-1\}$$

PROOF.–

It is assumed that the first and second statistical moments of the demand variable are known for each period *k*, that is, $E\{d(k)\}=\hat{d}_k$ and $V_{d(k)}=\sigma_d^2$ for each *k*.

The inventory variables S_k are statistically described by their means, $E\{S_k\}=\hat{S}_k$, as well as their variance, $V_{S_k}=E\left\{\left(S_k-\hat{S}_k\right)^2\right\}$.

The inventory balance equation [1.2] can be reformulated as:

$$\hat{S}_{k+1}=\hat{S}_k+U_{1,k}+\beta_2\cdot U_{2,k-\tau}+\delta.\hat{d}_{k-\tau'}-\hat{d}_k \text{ with } k\in\{0,1,\ldots,H-1\} \quad [1.17]$$

If we take the difference between equations [1.11] and [1.17], we obtain:

$$S_{k+1}-\hat{S}_{k+1}=\left(S_k-\hat{S}_k\right)+\left(\underbrace{U_{1,k}-\hat{U}_{1,k}}_{0}\right)+\left(\underbrace{\beta_2\cdot U_{2,k-\tau}-\beta_2\cdot\hat{U}_{2,k-\tau}}_{0}\right)+\delta\cdot\left(d_{k-\tau_1}-\hat{d}_{k-\tau_1}\right)-\left(d_k-\hat{d}_k\right)$$

$$\Rightarrow\left(S_{k+1}-\hat{S}_{k+1}\right)^2=\left(\left(S_k-\hat{S}_k\right)+\delta\cdot\left(d_{k-\tau_1}-\hat{d}_{k-\tau_1}\right)-\left(d_k-\hat{d}_k\right)\right)^2$$

$$\Rightarrow E\left\{\left(S_{k+1}-\hat{S}_{k+1}\right)^2\right\}=E\left\{\left(S_k-\hat{S}_k\right)^2\right\}+E\left\{(\delta)^2\left(\left(d_{k-\tau_1}-\hat{d}_{k-\tau_1}\right)\right)^2\right\}$$

$$+E\left\{\left(d_k-\hat{d}_k\right)^2\right\}+2\cdot E\left\{\left(S_k-\hat{S}_k\right)\cdot\delta\cdot\left(d_{k-\tau_1}-\hat{d}_{k-\tau_1}\right)\right\}$$

$$-2\cdot E\left\{\left(S_k-\hat{S}_k\right)\cdot\left(d_k-\hat{d}_k\right)\right\}-2\cdot E\left\{\left(d_k-\hat{d}_k\right)\cdot\delta\cdot\left(d_{k-\tau_1}-\hat{d}_{k-\tau_1}\right)\right\}$$

Since S_k and d_k are random independent variables, we can deduce that:

$$E\left\{\left(S_k-\hat{S}_k\right)\cdot\left(d_k-\hat{d}_k\right)\right\}=E\left\{S_k-\hat{S}_k\right\}E\left\{d_k-\hat{d}_k\right\}$$

$$E\left\{\left(S_k-\hat{S}_k\right)\cdot\delta\cdot\left(d_{k-\tau_1}-\hat{d}_{k-\tau_1}\right)\right\}=\delta\cdot E\left\{S_k-\hat{S}_k\right\}\cdot E\left\{d_{k-\tau_1}-\hat{d}_{k-\tau_1}\right\}$$

$$E\left\{\left(d_k-\hat{d}_k\right)\cdot\delta\cdot\left(d_{k-\tau_1}-\hat{d}_{k-\tau_1}\right)\right\}=\delta\cdot E\left\{d_k-\hat{d}_k\right\}\cdot E\left\{d_{k-\tau_1}-\hat{d}_{k-\tau_1}\right\}$$

Hence:

$$E\{S_k - \hat{S}_k\} = E\{S_k\} - E\{\hat{S}_k\} = 0$$

$$E\{d_k - \hat{d}_k\} = E\{d_k\} - E\{\hat{d}_k\} = 0$$

$$E\{d_{k-\tau_1} - \hat{d}_{k-\tau_1}\} = E\{d_{k-\tau_1}\} - E\{\hat{d}_{k-\tau_1}\} = 0$$

Therefore:

$$\Rightarrow E\left\{\left(S_{k+1} - \hat{S}_{k+1}\right)^2\right\} = E\left\{\left(S_k - \hat{S}_k\right)^2\right\} + (\delta)^2 \cdot E\left\{\left(d_{k-\tau_1} - \hat{d}_{k-\tau_1}\right)^2\right\} + E\left\{\left(d_k - \hat{d}_k\right)^2\right\}$$

If we assume that $V_S(k=0)=0$ and that σ_{d_k} is constant and equal to σ_d for all periods, we can deduce that:

$$V_s(k+1) = V_s(k) + \left((\delta)^2 + 1\right) \cdot \sigma_d^2$$

For:

$$k = 0, V_s(1) = \sigma_d^2,$$

$$V_s(k) = \sigma_d^2\left(k + (k-1) \cdot (\delta)^2\right)$$

Since $V_{S_k} = E\left\{\left(S_k - \hat{S}_k\right)^2\right\} = E\{S_k{}^2\} - \hat{S}_k{}^2$, we can write

$$E\{S_k{}^2\} - \hat{S}_k{}^2 = \sigma_d^2\left(k + (k-1) \cdot (\delta)^2\right)$$

Hence,

$$E\{S_k{}^2\} = \sigma_d^2\left(k + (k-1) \cdot (\delta)^2\right) + \hat{S}_k{}^2 \qquad [1.18]$$

Substituting [1.18] into [1.6], we obtain:

$$F(U_1,U_2)=\left(\sum_{k=0}^{H} f_k(U_{1,k},U_{2,k},S_k)=C_s\cdot E\{S_H^2\}+\sum_{k=0}^{H-1}\left[C_s\cdot E\{S_k^2\}+C_{pr1}\cdot U_{1,k}^2+C_{pr2}\cdot\beta_2\cdot U_{2,k-\tau}^2\right]\right)$$

$$F(U_1,U_2)=\sum_{k=0}^{H} C_s\times\left[\hat{S}_k^{\,2}+\left(k+(k-1)\cdot(\delta)^2\right)\cdot\sigma_d^2\right]+\sum_{k=1}^{H-1}\left(C_{pr1}\cdot U_{1,k}^2+C_{pr2}\cdot\beta_2\cdot U_{2,k-\tau}^2\right)$$

$$F(U_1,U_2)=\sum_{k=0}^{H} C_s\times\hat{S}_k^{\,2}+\sum_{k=0}^{H} C_s\times\left[\left(k+(k-1)\cdot\delta^2\right)\cdot\sigma_d^2\right]+\sum_{k=1}^{H-1}\left(C_{pr1}\cdot U_{1,k}^2+C_{pr2}\cdot\beta_2\cdot U_{2,k-\tau}^2\right)$$

$$F(U_1,U_2)=C_s\times\hat{S}_H^{\,2}+\sum_{k=1}^{H-1}\left(C_s\times\hat{S}_k^{\,2}+C_{pr1}\cdot U_{1,k}^2+C_{pr2}\cdot\beta_2\cdot U_{2,k-\tau}^2\right)+\sum_{k=0}^{H} C_s\times\left[\left(k+(k-1)\cdot\delta^2\right)\cdot\sigma_d^2\right]$$

$$F(U_1,U_2)=C_s\times\hat{S}_H^{\,2}+\sum_{k=1}^{H-1}\left(C_s\times\hat{S}_k^{\,2}+C_{pr1}\cdot U_{1,k}^2+C_{pr2}\cdot\beta_2\cdot U_{2,k-\tau}^2\right)+C_s\cdot\sigma_d^2\cdot\sum_{k=0}^{H} k+C_s\cdot\delta^2\cdot\sigma_d^2\cdot\sum_{k=0}^{H}(k-1)$$

$$F(U_1,U_2)=C_s\cdot\hat{S}_H^2+\sum_{k=0}^{H-1}\left[C_s\cdot\hat{S}_k^2+C_{pr1}\cdot U_{1,k}^2+C_{pr2}\cdot\beta_2\cdot U_{2,k-\tau}^2\right]+C_s\cdot\sigma_d^2\cdot\frac{(H+1)}{2}\cdot\left[H+\delta^2\cdot(H-2)\right]$$

1.3.3.3. *The service level constraint*

The transformation of the service level constraint into a deterministic form is given by Lemma 1.4.

LEMMA 1.4.–

$$\text{Prob}\left[S_{k+1}\geq 0\right]\geq\alpha\Rightarrow U_k\geq U_\alpha(S_k,\alpha)\ \text{ with } U_k=U_{1,k}+\beta_2\cdot U_{2,k-\tau}\quad[1.19]$$

where:

– $U_\alpha(S_k,\alpha)$ is the minimum cumulative production quantity;

– $U_\theta(S_k,\alpha)=(V_{d,k}\times V_{d,k-\tau'})\times\varphi^{-1}(\alpha)+\hat{d}_k-\delta\times\hat{d}_{k-\tau'}-S_k\quad k=0,1,\ldots,H-1$ is the variation in demand, *d*, during period *k*;

– $V_{d,k-\tau'}$ is the variation in demand, *d*, during period $k-\tau'$;

– φ is the cumulative Gaussian distribution function with mean $\left(\frac{1}{V_{d,k-\tau'}}\cdot\hat{d}_k - \frac{\delta}{V_{d,k}}\cdot\hat{d}_{k-\tau'}\right)$ and finite variance $\left(\left(\frac{1}{V_{d,k-\tau'}}\right)^2\cdot V_{d,k} + \left(-\frac{\delta}{V_{d,k}}\right)^2\cdot V_{d,k-\tau'} \geq 0\right)$;

– φ^{-1} is the inverse distribution function.

PROOF.–

$$S_{k+1} = S_k + U_{1,k} + \beta_2 \cdot U_{2,k-\tau} + r_k - d_k$$

$$\Pr ob\left(S_{k+1} \geq 0\right) \geq \alpha$$

$$\Pr ob\left(S_k + U_{1,k} + \beta_2 \cdot U_{2,k-\tau} + r_k - d_k \geq 0\right) \geq \alpha$$

$$\Pr ob\left(S_k + U_{1,k} + \beta_2 \cdot U_{2,k-\tau} + r_k \geq d_k\right) \geq \alpha$$

$$\Pr ob\left(S_k + U_{1,k} + \beta_2 \cdot U_{2,k-\tau} + r_k - \hat{d}_k \geq d_k - \hat{d}_k\right) \geq \alpha$$

$$\Pr ob\left(d_k - \hat{d}_k \leq S_k + U_{1,k} + \beta_2 \cdot U_{2,k-\tau} + r_k - \hat{d}_k\right) \geq \alpha$$

$$\Pr ob\left(d_k - \hat{d}_k \leq S_k + U_{1,k} + \beta_2 \cdot U_{2,k-\tau} + \delta.d_{k-\tau'} - \hat{d}_k\right) \geq \alpha$$

$$\Pr ob\left(d_k - \hat{d}_k - \delta\cdot d_{k-\tau'} + \delta\cdot\hat{d}_{k-\tau'} \leq S_k + U_{1,k} + \beta_2 \cdot U_{2,k-\tau} + \delta\cdot\hat{d}_{k-\tau'} - \hat{d}_k\right) \geq \alpha$$

$$\Pr ob\left(\frac{d_k - \hat{d}_k - \delta\cdot\left(d_{k-\tau'} - \hat{d}_{k-\tau'}\right)}{V_{d,k}\times V_{d,k-\tau'}} \leq \frac{S_k + U_{1,k} + \beta_2 \cdot U_{2,k-\tau} + \delta\cdot\hat{d}_{k-\tau'} - \hat{d}_k}{V_{d,k}\times V_{d,k-\tau'}}\right) \geq \alpha$$

$$\Pr ob\left(\frac{1}{V_{d,k-\tau'}}\times\frac{d_k - \hat{d}_k}{V_{d,k}} - \frac{\delta}{V_{d,k}}\times\frac{d_{k-\tau'} - \hat{d}_{k-\tau'}}{V_{d,k-\tau'}} \leq \frac{S_k + U_{1,k} + \beta_2 \cdot U_{2,k-\tau} + \delta\cdot\hat{d}_{k-\tau'} - \hat{d}_k}{V_{d,k}\times V_{d,k-\tau'}}\right) \geq \alpha$$

where:

– $\hat{d}_k$ is the mean demand during period k;

– $\hat{d}_{k-\tau'}$ is the mean demand during period ($k - \tau'$);

– $Var\left(d\left(k\right)\right) = V_{d,k} \geq 0$ is the variation in demand d during period k;

– $Var\left(d\left(k-\tau'\right)\right) = V_{d,k-\tau'} \geq 0$ is the variation in demand d during period ($k - \tau'$).

Note that $X = \left(\dfrac{d_k - \hat{d}_k}{V_{d,k}}\right)$ is a Gaussian random variable with a distribution identical to d_k and $Y = \left(\dfrac{d_{k-\tau'} - \hat{d}_{k-\tau'}}{V_{d,k-\tau'}}\right)$ is a Gaussian random variable with a distribution identical to $d_{k-\tau'}$. This formulation is equivalent to $\Pr ob(A \times X + B \times Y \leq C) \geq \theta$ with $A = \dfrac{1}{V_{d,k-\tau'}}$ and $B = -\dfrac{\delta}{V_{d,k}}$, $X' = A \times X$ is a random Gaussian variable with a distribution identical to $f_{X'} = \dfrac{1}{A} \times f\left(\dfrac{y}{A}\right)$, with mean $A \cdot \hat{d}_k = \dfrac{1}{V_{d,k-\tau'}} \cdot \hat{d}_k$ and variance $A^2 \cdot V_{d,k} = \left(\dfrac{1}{V_{d,k-\tau'}}\right)^2 \cdot V_{d,k} \geq 0$. It is a random Gaussian variable with a distribution identical to $f_{Y'} = -\dfrac{1}{B} \times f\left(\dfrac{y}{B}\right)$, with mean $B \cdot \hat{d}_{k-\tau'} = -\dfrac{\delta}{V_{d,k}} \cdot \hat{d}_{k-\tau'}$ and variance $B^2 \cdot V_{d,k-\tau'} = \left(-\dfrac{\delta}{V_{d,k}}\right)^2 \cdot V_{d,k-\tau'} \geq 0$.

Thus $T = X' + Y'$ is a random Gaussian variable with a distribution identical to $h = f_{X'} * f_{Y'}$, with mean $A \cdot \hat{d}_k + B \cdot \hat{d}_{k-\tau'}$ and variance $A^2 \cdot V_{d,k} + B^2 \cdot V_{d,k-\tau'} \geq 0$.

where φ is a probability distribution function of *T'*.

$$\varphi\left(\frac{S_k + U_{1,k} + \beta_2 \cdot U_{2,k-\tau} + \delta \cdot \hat{d}_{k-\tau'} - \hat{d}_k}{V_{d,k} \times V_{d,k-\tau'}}\right) \geq \alpha \qquad [1.20]$$

Since $\lim_{+\infty} \varphi \to 0$ and $\lim_{+\infty} \varphi \to 1$, we conclude that φ is strictly increasing. We note that φ is indefinitely differentiable, so we conclude that φ is invertible:

$$\frac{S_k + U_{1,k} + \beta_2 \cdot U_{2,k-\tau} + \delta \cdot \hat{d}_{k-\tau'} - \hat{d}_k}{V_{d,k} \times V_{d,k-\tau'}} \geq \varphi^{-1}(\alpha) \qquad [1.21]$$

$$U_{1,k} + \beta_2 \cdot U_{2,k-\tau} \geq \left(V_{d,k} \times V_{d,k-\tau'}\right) \times \varphi^{-1}(\alpha) + \hat{d}_k - \delta \times \hat{d}_{k-\tau'} - S_k$$

thus

$$\operatorname{Prob}\left(S_{k+1} \geq 0\right) \geq \alpha \Rightarrow \left(\left(U_{1,k} + \beta_2 \cdot U_{2,k-\tau}\right) \geq \left(V_{d,k} \times V_{d,k-\tau}\right) \times \varphi^{-1}(\alpha) + \hat{d}_k - \delta \times \hat{d}_{k-\tau'} - S_k\right) \quad k = 0, 1, \ldots, H-1$$

1.3.4. *Numerical example*

Using the same numerical example data as in the previous section (without retuned products case), a numerical example is presented in this section in order to illustrate the use of the analytical model developed in the case of production optimization with returned products. The following table provides a recap of the data.

Subcontractor machine data	
availability rate	$\beta_2 = 0.93$
Unit production cost	$C_{pr2} = 10$ *mu*
Maximal production rate	$U_2^{Max} = 130$ unit
Principal machine data	
Unit production cost	$C_{pr1} = 3$ mu/unit
Maximal production rate	$U_1^{Max} = 130$ unit
Length of production period	$\Delta t = 1$ month
Inventory data	
Inventory holding cost	Cs = 5 mu/Δt
Initial inventory level	S(0) = 200
Service level	$\alpha = 0.9$
Backordered product	
percentage of backordered product	$\delta \in [0, 0.5]$
returned production deadline	$\tau' = 1.\Delta t$

Table 1.9. *Numerical data*

The average demand with a standard deviation $\sigma = 2.23$ is presented in Table 1.10 below.

$d_0 = 150$	$d_1 = 170$	$d_2 = 150$	$d_3 = 150$	$d_4 = 150$	$d_5 = 140$	$d_6 = 160$	$d_7 = 140$
$d_8 = 160$	$d_9 = 130$	$d_{10} = 150$	$d_{11} = 140$	$d_{12} = 150$	$d_{13} = 120$	$d_{14} = 150$	$d_{15} = 130$
$d_{16} = 150$	$d_{17} = 110$	$d_{18} = 160$	$d_{19} = 130$	$d_{20} = 150$	$d_{21} = 120$	$d_{22} = 140$	$d_{23} = 160$

Table 1.10. *Mean demand*

Example 1

In this first example, we are interested in studying the influence of the returned product quantity on the optimal production plans for the main and subcontractor machines by varying the value of δ. In this case, we study the variability by considering that the transportation delay of subcontractor is fixed at $\tau = 1.\Delta t$. Figures 1.8 and 1.9 show an interesting result since the percentage products returned δ increases, so the principal and the subcontractor machines produce less, because the returned product quantity will be designed in the principal stock to help satisfy the customer and reduce the pressure on the principal and subcontractor machines.

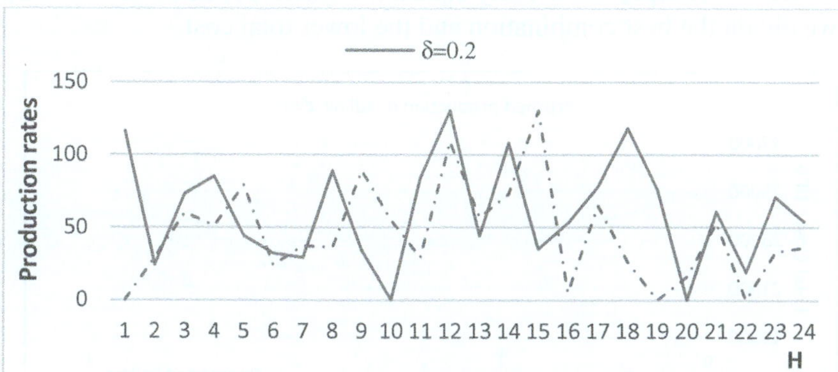

Figure 1.8. *Production rates of principal machine variation as a function of percentage of products returned δ with $\tau = 1.\Delta t$*

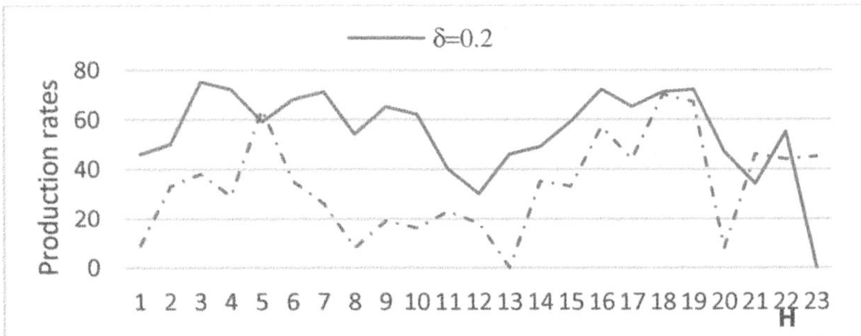

Figure 1.9. *Production rates of the subcontractor machine variation as a function of percentage of products returned δ with $\tau = 1.\Delta t$*

Example 2

In this example, we interpret from the variability of percentage of backordered products and the transportation delay of subcontracting. According to Figure 1.10, we note that the higher value of δ (percentage of backordered products) corresponds to the lower one of the optimal total cost. This can be explained by the fact that, when the returned product quantity increases, the main and subcontractor machines produce less and consequently the total production and inventory cost decrease. On the other hand, we note that in the case where the transportation delay equals the backorder delay ($\tau = \tau' = 1$), we obtain the best combination and the lower total cost.

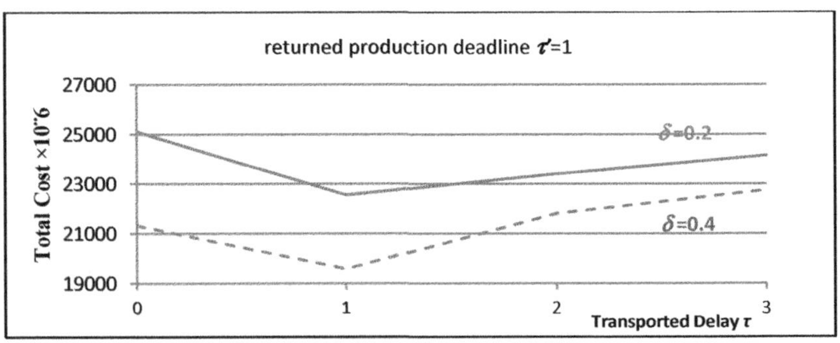

Figure 1.10. *Total cost variation as a function of the transportation delay τ and percentage of products returned δ*

1.3.5. *Optimization of returned products by a specified deadline*

In this section, we optimize the authorized deadline for product return. The main idea is to express a relationship between the subcontractor's transportation delay and the backorder delay. The relation is as follows $\tau' = \frac{\tau}{x}$ with $x \in \left]0, \tau\right]$. This relationship enables us to minimize the total production/inventory cost for the subcontractor transportation period selected. This section correlates the subcontractor transportation delay and the definition of the withdrawal (return) period.

The relationship between transportation delay and the deadline for returned products takes into account the minimization of the total production/inventory cost for a specified period of subcontractor transportation and correlates the delay relating to subcontractor transport with the definition of the withdrawal period. This relationship enables us to find the optimal combination of transportation delay and product return deadline to minimize cost.

As the deadline for returned products decreases, the quantity of products returned to the principal stockpile increases in order to satisfy customer demand and reduces pressure on the principal and subcontractor machines, and therefore decreases the transportation delay imposed by the use of a subcontractor. In this case, the subcontractor transport delay decreases, and consequently the total production/inventory cost decreases. In the opposite case, if the deadline for returned products increases, the quantity of products returned decreases and consequently the principal machine relies more heavily on the subcontracting machine. In this case, the transportation delay related to the use of a subcontractor decreases.

Two examples will be presented to highlight the application of the optimization idea; we consider the same numerical example data as in the previous section. We recall, that the objective is to determine the returned product deadline τ' in order to minimize the total production/inventory cost for different values of subcontractor transportation delay τ.

Example 1

We start with the first case where the subcontractor transportation delay is $\tau = 3$. In this case we vary the value of x in order to obtain the minimal total production and inventory cost. The above results illustrate the minimum total cost for different values of $x \in \left[1, \tau\right]$. For each value of x, we calculate the economical production plans of the subcontractor and the principal machines and the optimal returned production deadline τ'. Using Figure 1.11, we note that for the subcontractor transportation delay ($\tau = 3$, $x = 3$), the optimal total cost is equal to 2054200 mu obtained corresponding to an optimal returned product deadline $\tau' = 1$ and the economical production plans for the main and subcontractor machines are given by Tables 1.11 and 1.12. In the same way, Tables 1.13 and 1.14 show the result of the case 2 ($\tau = 3$, $x = 1.5$) where we obtain a minimal total production and inventory cost equals 2.04687×106 mu and an optimal returned production deadline $\tau' = 2$. Tables 1.15 and 1.16 present the result of case 3 ($\tau = 3$, $x = 1$) with a minimal cost equal to 2.25156×106 mu and optimal returned production deadline $\tau' = 3$.

Case 1: Transportation delay τ = 3, and x = 3

$\tau' = 1$ and $Cost = 2.0542\times10^{6}$ mu

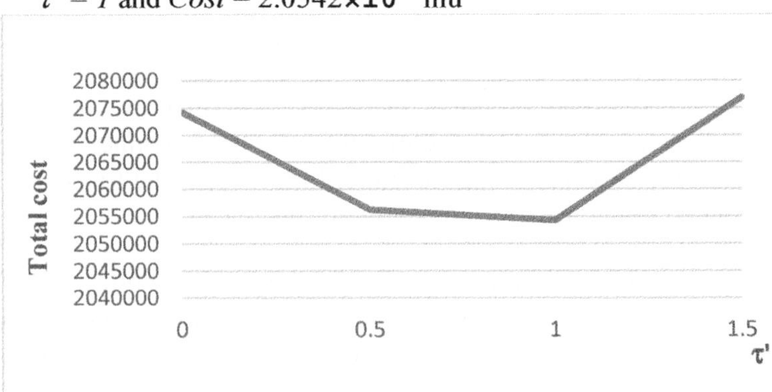

Figure 1.11. *Variability of returned production deadline with τ = 3, and x = 3*

$U_{1,0}$ = 83	$U_{1,1}$ = 35	$U_{1,2}$ = 102	$U_{1,3}$ = 42	$U_{1,4}$ = 102	$U_{1,5}$ = 106	$U_{1,6}$ = 120	$U_{1,7}$ = 0
$U_{1,8}$ = 40	$U_{1,9}$ = 103	$U_{1,10}$ = 49	$U_{1,11}$ = 56	$U_{1,12}$ = 89	$U_{1,13}$ = 100	$U_{1,14}$ = 40	$U_{1,15}$ = 20
$U_{1,16}$ = 28	$U_{1,17}$ = 16	$U_{1,18}$ = 85	$U_{1,19}$ = 1	$U_{1,20}$ = 63	$U_{1,21}$ = 0	$U_{1,22}$ = 99	$U_{1,23}$ = 116

Table 1.11. *Principal machine: $U^*_{1,k}$*

$U_{2,0}$ = 55	$U_{2,1}$ = 0	$U_{2,2}$ = 43	$U_{2,3}$ = 52	$U_{2,4}$ = 78	$U_{2,5}$ = 38	$U_{2,6}$ = 7	$U_{2,7}$ = 26
$U_{2,8}$ = 2	$U_{2,9}$ = 94	$U_{2,10}$ = 13	$U_{2,11}$ = 8	$U_{2,12}$ = 67	$U_{2,13}$ = 89	$U_{2,14}$ = 26	$U_{2,15}$ = 81
$U_{2,16}$ = 47	$U_{2,17}$ = 112	$U_{2,18}$ = 26	$U_{2,19}$ = 17	$U_{2,20}$ = 53	=	=	=

Table 1.12. *Subcontractor machine: $U^*_{2,k}$*

Case 2: Transportation delay τ = 3, and x = 1.5

$\tau' = 2$ and $Cost^* = 2.04687 \times 106$ mu

$U_{1,0}$ = 83	$U_{1,1}$ = 35	$U_{1,2}$ = 102	$U_{1,3}$ = 42	$U_{1,4}$ = 102	$U_{1,5}$ = 106	$U_{1,6}$ = 120	$U_{1,7}$ = 0
$U_{1,8}$ = 40	$U_{1,9}$ = 103	$U_{1,10}$ = 49	$U_{1,11}$ = 56	$U_{1,12}$ = 89	$U_{1,13}$ = 100	$U_{1,14}$ = 40	$U_{1,15}$ = 20
$U_{1,16}$ = 28	$U_{1,17}$ = 16	$U_{1,18}$ = 85	$U_{1,19}$ = 1	$U_{1,20}$ = 63	$U_{1,21}$ = 0	$U_{1,22}$ = 99	$U_{1,23}$ = 116

Table 1.13. *Principal machine: $U^*_{1,k}$*

$U_{2,0}$ = 55	$U_{2,1}$ = 0	$U_{2,2}$ = 43	$U_{2,3}$ = 52	$U_{2,4}$ = 78	$U_{2,5}$ = 38	$U_{2,6}$ = 7	$U_{2,7}$ = 26
$U_{2,8}$ = 2	$U_{2,9}$ = 94	$U_{2,10}$ = 13	$U_{2,11}$ = 8	$U_{2,12}$ = 67	$U_{2,13}$ = 89	$U_{2,14}$ = 26	$U_{2,15}$ = 81
$U_{2,16}$ = 47	$U_{2,17}$ = 112	$U_{2,18}$ = 26	$U_{2,19}$ = 17	$U_{2,20}$ = 53	=	=	=

Table 1.14. *Subcontractor machine: $U^*_{2,k}$*

Case 3: Transportation delay τ = 3, and x = 1

$\tau' = 3$ and Cost = 2.25156 × 106 mu

$U_{1,0}$ = 83	$U_{1,1}$ = 130	$U_{1,2}$ = 118	$U_{1,3}$ = 37	$U_{1,4}$ = 24	$U_{1,5}$ = 63	$U_{1,6}$ = 57	$U_{1,7}$ = 50
$U_{1,8}$ = 79	$U_{1,9}$ = 0	$U_{1,10}$ = 3	$U_{1,11}$ = 79	$U_{1,12}$ = 56	$U_{1,13}$ = 72	$U_{1,14}$ = 70	$U_{1,15}$ = 37
$U_{1,16}$ = 72	$U_{1,17}$ = 126	$U_{1,18}$ = 9	$U_{1,19}$ = 114	$U_{1,20}$ = 75	$U_{1,21}$ = 31	$U_{1,22}$ = 91	$U_{1,23}$ = 48

Table 1.15. *Principal machine: $U^*_{1,k}$*

$U_{2,0}$ = 50	$U_{2,1}$ = 92	$U_{2,2}$ = 39	$U_{2,3}$ = 95	$U_{2,4}$ = 52	$U_{2,5}$ = 14	$U_{2,6}$ = 78	$U_{2,7}$ = 82
$U_{2,8}$ = 75	$U_{2,9}$ = 26	$U_{2,10}$ = 8	$U_{2,11}$ = 37	$U_{2,12}$ = 59	$U_{2,13}$ = 12	$U_{2,14}$ = 31	$U_{2,15}$ = 20
$U_{2,16}$ = 49	$U_{2,17}$ = 87	$U_{2,18}$ = 0	$U_{2,19}$ = 15	$U_{2,20}$ = 111	=	=	=

Table 1.16. *Subcontractor machine: $U^*_{2,k}$*

Example 2

In this second example, we vary the value of the subcontractor transportation delay τ ($\tau = 1$ and $\tau = 3$) and we compare the different results such as the optimal subcontractor transportation delay, the optimal total cost and the optimal returned product deadline. Using Figure 1.12, we note that for the subcontractor transportation delay $\tau = 1$, the optimal total cost is obtained for $x = 1$ corresponding to the optimal returned product deadline $\tau' = 1$ as for subcontractor transportation delay $\tau = 3$ the withdrawal right requires that the optimal returned product deadline is $\tau' = 2$.

Hence, based on the results of the previous sections as well as this section, we note that when τ increases, we produce more on the principal machine and store more to meet future periods. According to the previous results presented through the variability of τ, the storage costs are visibly impacted. Thus, the desirable strategy to adopt for the return of products from customers is finding an optimal period (maximal delay of backorder) in order to remove pressure on the stock and to reduce costs, therefore the choice of τ' must be optimized and correlated with the subcontractor transportation delay τ. The results presented in this section show the existence of an optimum.

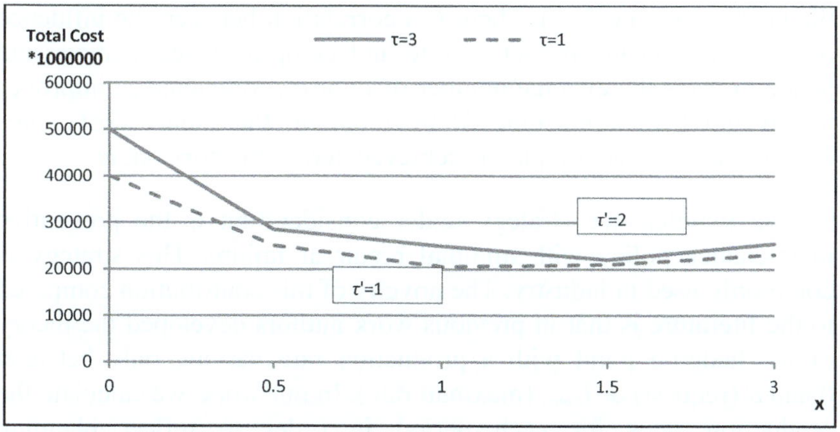

Figure 1.12. *Variability of returned production deadline*

1.4. Joint maintenance policy

The aim of this section is to determine an effective joint maintenance policy with an optimal production plan that integrates the various constraints related to the production rates, transportation delay and product return deadline. The maintenance policy has to ensure the creation of the volumes necessary to meet customer demand. We adopt a maintenance policy based on block strategy and combine it with demand forecast. This maintenance policy ensures the reliability necessary for the main machine, which is directly impacted by the delivery deadline, as demonstrated previously. Its advantage is that it is easy to apply without disrupting the production plan. In this section, we optimize this proposed policy in order to determine the optimal period on the horizon, H, according to a cost minimization criterion.

1.4.1. *Description of the problem*

Machine M is subject to random failures. Its probability of degradation is described by the probability density function of time to failure $f(t)$, for which the failure rate $\lambda(t)$ increases with both time and the production rate $U_l(t)$. There is a correlation between the influence of the variation in production rate and equipment degradation, and hence impacts the average number of failures. Maintenance improves the availability of machine M by reducing the failure costs, thus ensuring the production plan is achieved across the horizon, H.

The maintenance strategy under consideration is the preventive maintenance policy with minimal repair at failure. This strategy is commonly used in industry. The novelty of this contribution compared to the literature is that in previous work authors developed the theory of the hedging point with a production rate varying only between 0 and d (request) or U_{max} (maximal rate). In our work, we calculate the production rates for each period in order to reflect planning commitments while correlating the degradation of machine M's production rates. We optimize the production rate over a finite horizon. As production rates vary from one period to another, the maintenance plan must change accordingly, because degradation is not uniform over the planning horizon. The production rate is used to calculate the periodicities of preventive maintenance actions.

The horizon, H, is partitioned into N equal parts, each of length T. Perfect preventive maintenance or replacement is performed periodically at times $i.T$, $i = 0,1,\ldots,N$ and $N.T = H.\Delta t$, following which the unit is as good as new. When a unit fails between preventive maintenance actions, only minimal repair is performed. It is assumed that the repair and replacement times are negligible.

The dependence of system degradation on the production plan is manifested by an increased failure rate according to both increased time and production rate [HAJ 09]. That is why we focus on the joint optimization strategy in which we consider maintenance in order to establish the optimal maintenance strategy according to a criterion of cost minimization characterized by the optimal number, N^*, of preventive maintenance actions to be performed over the finite horizon, $H.\Delta t$.

1.4.2. *Analytical study*

The analytical expression of the total maintenance cost is given in equation [1.22]:

$$\xi_M\left(U_1,N\right)=C_{pm}\times(N-1)+C_{cm}\times A_M\left(U_1,N\right) \qquad [1.22]$$

where $N\in\{1,2,3.....\}$ and $A_M\left(U_1,N\right)$ correspond to the expected number of failures that occur during horizon *H*, considering the production rate in each production period Δt.

Each period, *k*, of the horizon, *H*, is characterized by its own production rate, U_k, which is established from the production plan. The failure rate evolves during each interval according to the production rate adopted in this interval. It also depends on the cumulative effect of the failure at the end of the previous period. As per the approach taken by Hajej *et al.* [HAJ 09], degradation at the end of the period is accounted for. In fact, the failure rate in interval *k* is expressed as follows:

$$\lambda_k(t)=\lambda_{k-1}(\Delta t)+\frac{U_{1,k}}{U_{\max}}\lambda_n(t) \qquad \forall t\in\left[0,\Delta t\right] \qquad [1.23]$$

with

$$\lambda_{k=0}=\lambda_0 \text{ and } \Delta\lambda_k(t)=\frac{U_{1,k}}{U_{\max}}\lambda_n(t)$$

In equation [1.23], $\lambda_n(t)$ is the nominal failure rate corresponding to the maximal production rate.

We recall that Hajej *et al.* [HAJ 09] assumed that machine degradation is linear according to the production rate. We can write the failure rate function as expressed by Hajej *et al.* [HAJ 09] in the following way:

$$\lambda_{k,j}(t) = \lambda_{0,j} + \sum_{l=1}^{k-1} \frac{U_{1,l,j}}{U_{\max}} \lambda_n(\Delta t) + \frac{U_{1,k,j}}{U_{\max}} \lambda_n(t) \ , \ t \in [0, \Delta t] \qquad [1.24]$$

Let *In* denote the integer part of (.). The average number of failures over horizon $H.\Delta t$ is thus:

$$\begin{aligned} A_M(U,N) = \sum_{j=0}^{N-1} \Bigg[& \sum_{i=In(j\times\frac{T}{\Delta t})+1}^{In\left((j+1)\times\frac{T}{\Delta t}\right)} \int_0^{\Delta t} \lambda_{i,j}(t) + \int_0^{(j+1)\times T - In\left((j+1)\times\frac{T}{\Delta t}\right)\times\Delta t} \lambda_{In\left((j+1)\times\frac{T}{\Delta t}\right)+1,j}(t)dt \\ & + \int_0^{\left(In\left((j+1)\times\frac{T}{\Delta t}\right)+1\right)\times\Delta t-(j+1)\times T} \frac{U_{1,\left(In\left((j+1)\times\frac{T}{\Delta t}\right)+1\right)}}{U_{\max}} \times \lambda_n(t)dt \Bigg] \end{aligned} \qquad [1.25]$$

In expression [1.25] we therefore replace $\lambda_i(t)$:

$$A_M(U,T) = \sum_{j=0}^{N-1} \left(\begin{aligned} & \left(In\left((j+1)\times\frac{T}{\Delta t}\right) - In\left(j\times\frac{T}{\Delta t}\right)\right)\times\Delta t\times\lambda_{0,j}(t_0) + \\ & + \sum_{i=In\left(j\times\frac{T}{\Delta t}\right)+1}^{In\left((j+1)\times\frac{T}{\Delta t}\right)} \int_0^{\Delta t} \left(\sum_{l=In\left(j\times\frac{T}{\Delta t}\right)+1}^{i-1} \frac{U_{1,l,j}}{U_{\max}} \cdot \lambda_n(\Delta t) \right) dt \\ & + \sum_{i=In\left(j\times\frac{T}{\Delta t}\right)+1}^{In\left((j+1)\times\frac{T}{\Delta t}\right)} \int_0^{\Delta t} \frac{U_{1,i,j}}{U_{\max}} \cdot \lambda_n(t)dt + \\ & \int_0^{(j+1)\times T - In\left(\frac{(j+1)\times T}{\Delta t}\right)\times\Delta t} \left(\sum_{l=In\left(j\times\frac{T}{\Delta t}\right)+1}^{In\left(\frac{(j+1)\times T}{\Delta t}\right)} \frac{U_{1,l,j}}{U_{\max}} \cdot \lambda_n(\Delta t) \right) dt \\ & + \int_0^{(j+1)\times T - In\left(\frac{(j+1)\times T}{\Delta t}\right)\times\Delta t} \frac{U_{1,\left(In\left(\frac{(j+1)\times T}{\Delta t}\right)+1\right),(j+1)}}{U_{\max}} \cdot \lambda_n(t)dt \\ & + \frac{U_{1,\left(In\left(\frac{(j+1)\cdot T}{\Delta t}\right)+1\right),(j+1)}}{U_{\max}} \times \int_0^{\left(In\left((j+1)\cdot\times\frac{T}{\Delta t}\right)+1\right)\times\Delta t-(j+1)T} \lambda_n(t)\,dt \end{aligned} \right)$$

We now replace $T = H/N$:

$$A_M\left(U, {}^{H}\!/_{N}\right) = \sum_{j=0}^{N-1} \left(\begin{array}{l} \left(In\left((j+1)\times\frac{H}{N\cdot\Delta t}\right) - In\left(j\times\frac{H}{N\cdot\Delta t}\right)\right)\times\Delta t\times\lambda_0(t_0) + \\ \frac{\lambda_0(\Delta t)\times\Delta t}{U_{\max}}\times\sum_{i=In\left(j\times\frac{T}{\Delta t}\right)+1}^{In\left((j+1)\times\frac{H}{N\cdot\Delta t}\right)}\sum_{l=1}^{i-1}U_{1,l}dt + \frac{1}{U_{\max}}\cdot\sum_{i=In\left(j\times\frac{H}{N\cdot\Delta t}\right)+1}^{In\left((j+1)\times\frac{H}{N\cdot\Delta t}\right)}\int_0^{\Delta t}U_{1,i}\cdot\lambda_n(t)dt + \\ \sum_{l=1}^{In\left(\frac{(j+1)\times H}{N\cdot\Delta t}\right)}\frac{U_{1,l}}{U_{\max}}\cdot\lambda_n(\Delta t)\cdot\left((j+1)\times\frac{H}{N} - In\left(\frac{(j+1)\times H}{N\cdot\Delta t}\right)\times\Delta t\right) \\ + \int_0^{(j+1)\times\frac{H}{N}-In\left(\frac{(j+1)\times H}{N\cdot\Delta t}\right)\times\Delta t}\frac{U_{1,\left(In\left(\frac{(j+1)\times H}{N\cdot\Delta t}\right)+1\right),(j+1)}}{U_{\max}}\cdot\lambda_n(t)dt \\ + \frac{U_{1,\left(In\left(\frac{(j+1)\cdot H}{N\cdot\Delta t}\right)+1\right),(j+1)}}{U_{\max}}\times\int_0^{\left(In\left((j+1)\times\frac{H}{N}\right)+1\right)\times\Delta t-(j+1)\times\frac{H}{N}\times\Delta t}\lambda_n(t)\,dt \end{array} \right)$$

1.4.3. *Numerical example*

The following numerical example is considered to illustrate our approach. We use the same data for production, inventory and service level of the previous numerical example. The number H of periods Δt is equal to 24, with $\Delta t = 1$. The machine M_l has a degradation law characterized by a Weibull distribution. Recall that our contribution for maintenance is to study the influence of the variation of the production rate on the machine degradation that is new in the literature. Equation [1.24] shows the evolution of the machine failure rate according to its use (which in our case is the production rate for each period) respecting at the same time the continuity of the equipment reliability for a period to another. This equation has been validated in other scientific papers. Concerning the Weibull law, we have chosen a numerical example where we have assumed that the degradation of the equipment follows the Weibull law with parameters $\gamma = 2$ and $\beta = 100$ (with these two parameters, the degradation is linear $\gamma = 2$). From this equation, we determined the average number of failures assuming that after each preventive maintenance action the equipment is on state "as good as new" and that maintenance action may be applied during the production as it can be at the end of the period.

The other data are as follows: $C_{pm} = 500$, $C_{cm} = 4500$, $\Delta t = 1$month, $\lambda_0 = 0.2$, $C_{pr1} = 3mu$, $C_{pr2} = 10mu$, $U_1^{Max} = 130$, $U_2^{Max} = 130$, $\beta_2 = 0.93$, service level $\alpha = 0.95$, $Cs = 5mu$, initial inventory $S_0 = 200$ and the standard deviation $\sigma = 2.23$.

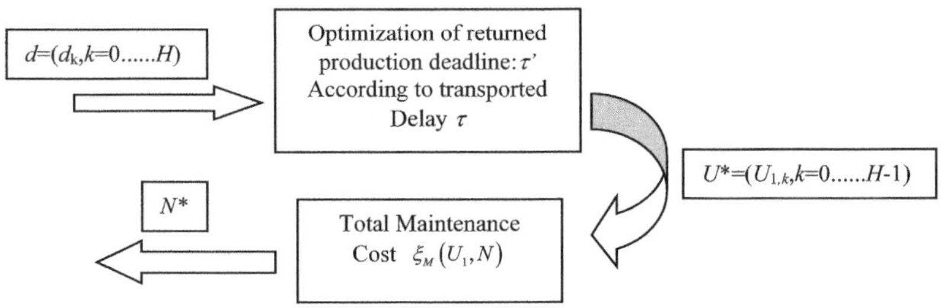

Figure 1.13. *Maintenance optimization*

Using the previous sections, an optimization procedure was implemented using the Numerical Algorithms for Constrained Global Optimization (Nelder Mead methods) with MATHEMATICA.

In this section, we are interested in different examples where we determine the optimal strategy of maintenance characterized by the optimal number of preventive maintenance actions N* in the case with returned products and the case without returned products.

Example 1 (without returned products)

According to the economical production plan for the main machine obtained in the case without returned products with different values of subcontracting transportation delay τ and the relationship of failure rate presented in equation [1.24], we determine the optimal number of preventive maintenance actions N^* for each value of τ ($\tau = 0$, $\tau = 1$, $\tau = 4$) .

Figures 1.14–1.16 illustrate the minimum total cost for different values of the number, N, of preventive maintenance actions to be performed. We note that for the subcontractor transportation delay $\tau = 0$ and according to the economical production plan for the main machine obtained in Table 1.3, we obtained an optimal total cost equal to 5059700.2 *mu,* with two preventive maintenance actions being performed during the horizon $H = 24$. This optimal total cost becomes more expensive for $\tau = 1$ with the economical production plan that is presented in Table 1.5, and equal to 5229300 *mu* with three preventive maintenance actions, and significantly more expensive equal to 6251100.1 *mu* with six preventive maintenance actions being performed for subcontracting transportation delay $\tau = 4$.

Optimal number N^* of preventive maintenance actions

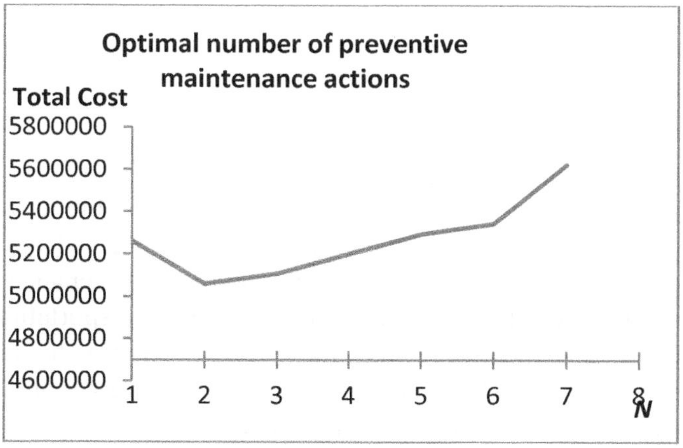

Figure 1.14. *Total cost variation as a function of N with $\tau = 0$*

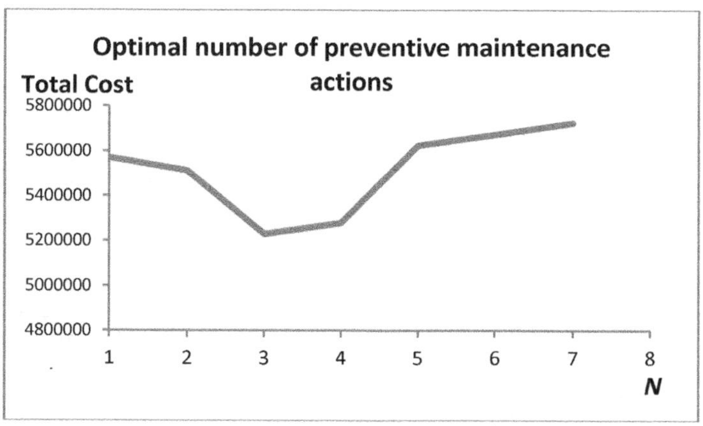

Figure 1.15. *Total cost variation as a function of N with $\tau = 1$*

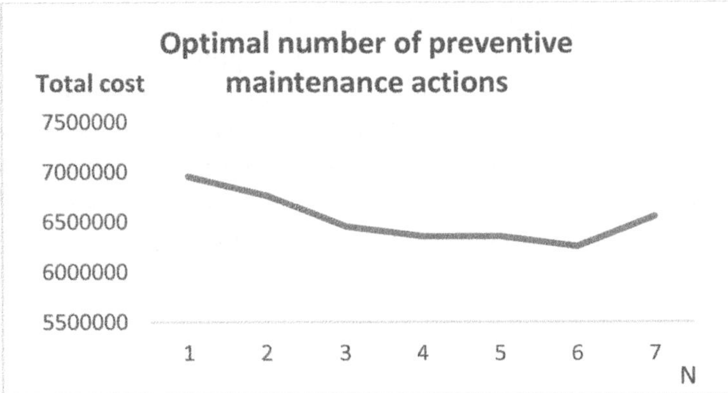

Figure 1.16. *Total cost variation as a function of N with $\tau = 4$*

Hence, from Figure 1.17, we can see that the higher value of transportation delay τ corresponds to the higher value of the optimal total cost and the optimal number of preventive maintenance actions. This can be explained by the fact that, as transportation delay τ increases, the principal machine produces more to meet the customers' demands, thus the machine will undergo more failures. According to the previous results presented through the variability of τ, subcontractor transport delay is visibly impacted.

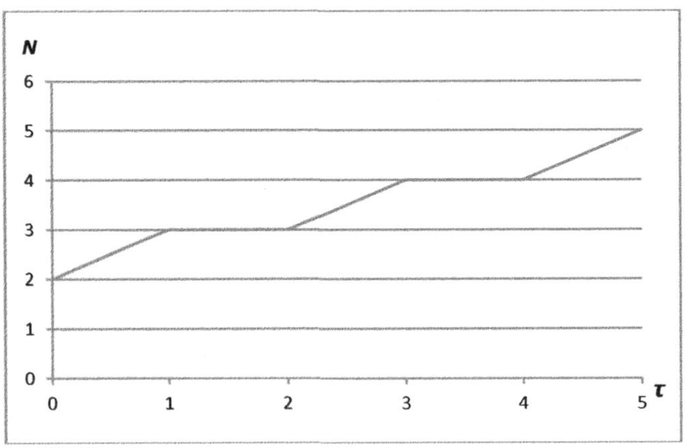

Figure 1.17. *MP number variation as a function of τ*

Example 2 (with returned products)

We are interested in this example by the case where we consider the products returned by the production problem. In this case, we optimize the maintenance strategy according to the result obtained in example 1 of section 3.4. In this case, we determine the optimal number of preventive maintenance actions N^* corresponding to the optimal production plan for the main machine with a percentage of returned products $\delta = 0.2$ and a transportation delay of subcontractor fixed at $\tau = 1.\Delta t$. Figure 1.18 shows that the optimal number of preventive maintenance actions $N^* = 2$ and consequently two preventive maintenance actions are performed during the horizon $H = 24$. The last result corresponding to the optimal number of preventive maintenance actions ($N^* = 2$) is lower than the result of example 1 of the case without returned products $N^* = 3$. We can see that the higher values of returned products corresponds to the lower values of production rates. Consequently, the plan of production which has a lower value of production rates is characterized by a lower optimal number of preventive maintenance actions $N^* = 2$. This can be explained by the fact that, as production rates decrease, normally the failure rate decreases as well as the average number of failures and consequently the number of preventive maintenance actions.

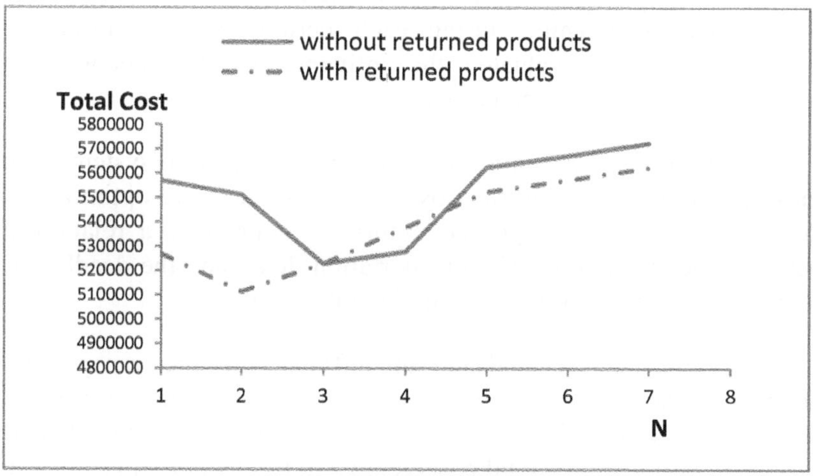

Figure 1.18. *Total cost variation as a function of N with $\tau = 1$*

1.5. Conclusion

This chapter analyzes a constrained stochastic production and maintenance planning problem. In order to meet demand across the horizon considered, subcontracting is required. At a given service level, we have formulated and solved the related stochastic production/inventory and maintenance problem.

The subcontractor, as it is based elsewhere, introduces a delay as stock needs to be transported to the principal stockpile. It also introduces constraints related to availability, maximum production capacity and unit production cost. The key is to consider the influence of the transportation delay introduced by the subcontractor in the context of production, inventory and maintenance.

In the initial case (section 1.2), where we considered a production system without returned products, we formulated and solved a linear quadratic stochastic production problem to obtain a preliminary production plan that took into account the transportation delay introduced by a subcontractor. Using the Modigliani–Muth–Simon model, the preliminary production plan included the production and inventory costs at a given service level. This plan defined the production rates for the main production system and for the subcontractor's machine during each period over the production horizon. We then evaluated the influence of the subcontracting transportation delay on the production plan.

In section 1.3, we went on to consider a production system taking into account products returned by customers. Here, we optimized the deadline authorized for product return by expressing a relationship between the subcontractor's transportation delay and the deadline for returned products in order to minimize total cost.

In section 1.4, we analyzed the influence of the transportation delay introduced by the subcontractor in the maintenance policy in order to determine the optimal number of preventive maintenance actions to minimize the total costs related to production, inventory and maintenance.

In the future, we will consider a more complex system with a subcontractor transportation cost. We will also take into account the cost of executing an order. For the maintenance strategy, we will consider the new hypothesis that the corrective and preventive times are not negligible. In this case we will deal with new policy to recover the quantity lost during maintenance.

In other future research, we will consider the aspect of leasing in our problem. The idea is to switch between a subcontracting machine and leased machines in order to minimize the production, inventory and maintenance costs.

2

Sequentially Optimizing Production, Maintenance and Delivery Activities Taking into Account Product Returns

This chapter develops and analyzes a stochastic optimization problem with a service level constraint to generate a sequentially optimal plan of production, maintenance and delivery activities in a deteriorating manufacturing system. Stochastic demand and product returns are both accounted for, the latter of which allows for re-stocking new products returned by the customer that are thus in a saleable condition. A constrained production–maintenance–delivery problem that incorporates service level, stochastic demand, delivery time, failure rate and proportion of products returned is formulated based on the quadratic model. This quadratic formulation is adapted to provide inventory, delivery, production and maintenance policies. The objective of this chapter is to study the influence of delivery time on the planning of production, maintenance and delivery activities. Finally, we work through a simulation to illustrate the exploitation of the proposed approach.

2.1. Introduction

In recent decades companies have had to cope with decreasing profit margins due to the competitive environment, which in turn has motivated them to seek improvements in their production and maintenance planning performance. In the literature there are many papers that have treated the maintenance planning problem independently of the production planning problem, despite the fact that in the past few decades we have seen the emergence of research that treats both problems simultaneously. Holt *et al.* [HOL 60] have developed a linear decision rule that considers inventory and production planning without maintenance, and which provides an important contribution to the literature. The proposed analytical rule allows us to obtain the optimal solution to the quadratic cost function of production, inventory and workforce levels. The principle of this method is to minimize the inventory and workforce equations.

According to Bertesekas [BER 95], the Modigliani–Muth–Simon model is one of the first to deal with the certainty equivalence principle for dynamic linear quadratic problems. This model is usually used as a benchmarking tool in order to compare different production planning approaches and to provide managers and decision-makers with perspectives on and ideas about how to manage a firm's material resources [SIN 96].

Hax and Candea [HAX 84], for example, proved that the Modigliani–Muth–Simon quadratic approach is useful for evaluating the production process. For example, by using the Modigliani–Muth–Simon model, the quadratic inventory cost describes and takes into account the possible status of inventory, whether negative (shortage) or positive (overstocking). In this context, Buzacott [BUZ 67] analyzed the role of a buffer inventory in increasing the productivity of an unreliable production system.

Aghezzaf *et al.* [AGH 07] developed a sequential maintenance and production planning model for a production system subject to random failures. This model takes into account the system reliability parameters and capacity in the development of the optimal production plan.

Van der Duyn Schouten and Vanneste [VAN 95a] introduced a finite-state Markov decision model for the optimal preventive maintenance of an installation in a production line with an intermediate buffer. Under certain suitable conditions they proved that, for each fixed buffer level, the optimal policy is of the control-limit type.

Das and Sarkar [DAS 99] considered a production–inventory system, where inventory is maintained according to an (s, S) policy and the production process is subject to failure. The authors studied a preventive maintenance policy and employed a simulation-based optimization algorithm to obtain optimal policy parameters.

Iravani and Duenyas [IRA 02] considered a make-to-stock production/inventory system consisting of a single deteriorating machine that produces a single item. They formulated the integrated decisions of maintenance and production using a Markov decision process.

Rezg *et al.* [REZ 04], as well as Chelbi and Ait-Kadi [CHE 04], analyzed a strategy that builds a safety inventory in order to satisfy demand when production is interrupted during maintenance actions. Cheung and Hausmann [CHE 97] developed the sequential optimization of a strategic safety inventory and age-based maintenance policy (taking into account the age of machines, as machines need more maintenance as they age). In practice the failure rate increases with both the equipment use and time, but the latter is rarely considered in the literature. For instance, Hu *et al.* [HU 94] analyzed the optimal conditions for a hedging point strategy controlling a production system that takes into account machine failure that depends on the number of parts that are produced by the machine.

Most works in the literature consider a perfect service level and perfect manufacturing systems, and do not take into account the impact of service level and the proportion of defective items on system performance measures and costs. More recently, Hajej *et al.* [HAJ 09, HAJ 11] considered a manufacturing system composed of a randomly failing machine that satisfies a stochastic demand under a service level constraint. The authors ignored some crucial properties of a production system, however, such as transportation in terms of delivery time and the quantity transported.

Most manufacturers are trying to decrease the delivery time, i.e. the time the parts spend travelling between the store and the customer's warehouse. This delivery time has a great impact on system performance. Turki *et al.* [TUR 12a, TUR 12b] have taken into account random demands and machine failures. These authors studied the influence of delivery time on optimal inventory level. Dolgui and Ould-Louly [DOL 02] developed a supply planning model based on a specified stochastic lead time and proposed a method for determining the optimal planned lead time. Lee [LEE 05] proposed a model for evaluating investment strategies in preventive maintenance and inventory in an imperfect manufacturing system taking delivery time into consideration.

The quantity of products transported, a property of manufacturing systems that has been considered recently, relates to the quantity of products transported from the stockpile to the warehouse [FUN 05]. In order to achieve a targeted service level, a warehouse needs to be purchased to contain enough parts. This depends to a certain degree upon the quantity being transferred between the manufacturer and the customer warehouse.

The study of combined production, maintenance and delivery strategies is a very recent topic in the literature. Hajej *et al.* [HAJ 14a] have determined optimal production plans for principal and subcontracting machines that minimize the total production and inventory cost in systems with and without returned products that meet a specified service level and incorporate subcontracting transportation delay. In Turki *et al.* [TUR 12b] a manufacturing model was considered that takes into account the transportation delay between the principal stockpile and customer warehouse. They then studied the impact of delivery time on maintenance planning in a simple case. Motivated by this work, we decided to combine production, maintenance and delivery plans taking into consideration machine failures, stochastic demand, delivery time, service level and returned products (where the products previously delivered to the customer are returned in a good-as-new state and are therefore in a saleable condition). Hence, this work is original as it studies the impact of delivery time, the rate at which products are returned and service level on optimal production and maintenance plans, taking into account delivery. We make three contributions in this chapter:

– we consider the relationship between transportation delay and the rate at which products are returned, and then study their influence on the optimal production, delivery and maintenance plans;

– we add to the literature about maintenance and the contribution of the influence of variation in production to machine degradation. In our study we take into account the evolution of machine failure rate according to the production rate for each period, respecting the continuity of equipment reliability from one period to another;

– we study the interaction between:

- the production rate,
- the quantity of products transported,
- the delay in delivery,
- the amount of products returned,
- the service level.

The proposed approach is based on sequential resolution. We determine an optimal production plan and as a consequence determine the optimal maintenance policy.

Section 2.2 presents a stochastic delivery, production and maintenance model considering the impact of delivery time, while section 2.3 transforms this model in to a deterministic equivalent. Numerical examples are presented in section 2.4. Section 2.5 concludes the chapter.

2.2. Planning of production, delivery and maintenance

This section presents the notations (section 2.2.1) and assumptions (section 2.2.2) used throughout this chapter. It goes on to explain the model (section 2.2.3) and the derivation of cost expressions (section 2.2.4).

2.2.1. *Notation*

The following parameters are used in the explanation of the model:

δ	rate at which products are sent back to the warehouse S_0 (where the products are still in a saleable condition)
τ	time between the customer receiving the products and returning them (at a rate δ) to the warehouse S_0
$r(k)$	quantity of products returned by the customer after τ_r
Δt	length of the production period
τ:	delivery time (the value of the delivery time is a multiple of Δt)
H	number of production periods in the planning horizon
$H.\Delta t$	the length of the finite planning horizon
$u(k)$	the number of products produced by machine M during period k, with: $k = \{0, 1,\ldots, H-1\}$ $U = \{u(0), u(1), \ldots, u(H-1)\}$
$Q(k)$	the number of products transported from S_1 to S_0 during period k, with: $k = \{0, 1,\ldots, H-1\}$ $Q = \{Q(0), Q(1), \ldots, Q(H-1)\}$
$\hat{d}(k)$	average demand during period k where $k = \{0, 1,\ldots, H\}$
V_k	variation in demand during period k where $k = \{0, 1,\ldots, H\}$
$S_1(k)$	inventory level at S_1 at the end of period k where $k = \{0, 1,\ldots, H\}$
$\hat{S}_1(k)$	average inventory level at S_1 during period k where $k = \{0, 1,\ldots, H\}$

$S_0(k)$	inventory level at S_0 at the end of period k where $k = \{0, 1,\ldots, H\}$
$\hat{S}_0(k)$	average inventory level at S_0 during period k where $k = \{0, 1,\ldots, H\}$
C_{pr}	unit production cost of machine M
Cs_1	inventory holding cost of one product unit during one period at S_1
Cs_0	inventory holding cost of one product unit during one period at S_0
C_l	delivery cost
Q_v	delivery vehicle capacity
C_{TM}	total maintenance cost
$G_{USQ}(k)$	expected quadratic value of the sum of the inventory, production and delivery costs for period k
C_{pm}	preventive maintenance action cost
C_{cm}	corrective maintenance action cost
mu	monetary unit
U_{max}	maximum production rate of machine M
U_{min}	minimum production rate of machine M
θ	probability index related to customer satisfaction and expressing the service level
$F(t)$	probability distribution function associated with the time to failure of M
$R(t)$	reliability function, equal to $1 - F(t)$
$\lambda_n(t)$	nominal failure rate corresponding to the maximal production rate
$\lambda_k(t)$	machine failure rate during period k where $k = \{0, 1,\ldots, H\}$

S_1^0	initial inventory level of S_1
S_0^0	initial inventory level of S_0
N	number of preventive maintenance actions over the finite horizon, H
$\varphi_M(U,N)$	average number of failures over the finite horizon, H
φ	cumulative Gaussian distribution function

$u(k)$, $Q(k)$ and N decision variables

In this section, a discrete-time model for sequentially optimizing production, maintenance and delivery planning is formulated in a sequential way over a finite horizon. In this chapter, we consider a one-machine manufacturing system that produces one type of product and which is composed of a customer with stochastic demand and two inventories (S_1, S_0). S_1 is the manufacturing store in which the manufactured products are initially kept. These products are then transported to a second store (S_0) for sale. S_0 is thus the purchase warehouse in which products taken from S_1 are stored and from where customer demand is satisfied.

It takes time to transport the manufactured products from the manufacturing store S_1 to the purchase warehouse S_0. Thus, we consider a delivery time τ between S_1 and S_0 (see Figure 2.1). In order to propose a more realistic system, we consider that the manufactured parts are transported from S_1 to S_0 by a conveyance that has the capacity Q_v. Warehouse S_0 allows customer demand to be satisfied according to a prescribed service level, θ.

As we mentioned in the introduction, a proportion of the manufactured products delivered to the customer is eventually returned to S_0. Machine M is subject to random failures, and the failure rate varies according to both time and to the number of products produced by machine M.

The goal is to simultaneously provide optimal delivery, maintenance and production plans that satisfy the stochastic demand in a sequential manner. The objective is to minimize the sum of the manufacturing, delivery, inventory and maintenance costs.

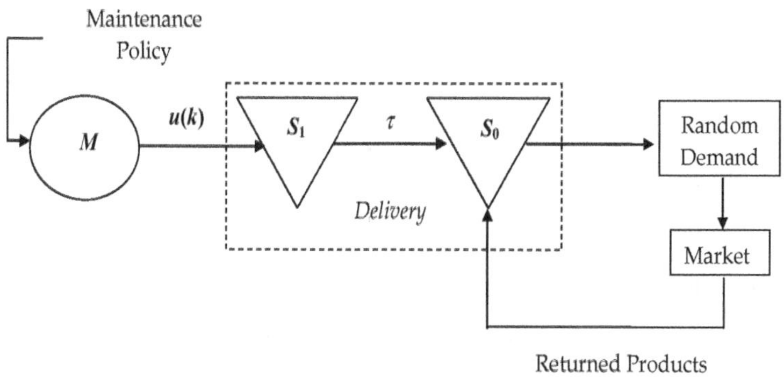

Figure 2.1. *Description of the problem*

2.2.2. *Context and assumptions*

In this model, we make the following assumptions:

1) customer demand follows Gaussian distribution and every unsatisfied demand in a period is backordered;

2) the delivery time τ is constant and is a multiple of Δt (the length of a period). If we assume that the products leave S_1 in period k, they will arrive at S_0 in period $k+\tau/\Delta t$;

3) the returned products are still new and in saleable condition;

4) the rate at which products are returned (δ) is constant;

5) the return deadline (τ_r) is constant;

6) the quantity of products returned by the customer ($r(k)$) is proportional to demand and products should be returned to S_0 after the specific deadline τ_r. This proportion is specified by the rate at which products are returned (δ);

7) the maximum (U_{max}) and minimum (U_{min}) production rates are constant and known;

8) the vehicle capacity (Q_v) is constant and known;

9) we assume that after each preventive maintenance action, the equipment is "as good as new" and that maintenance actions may be applied during production as well as at the end of the period;

10) we assume that the durations for repair and overhaul are negligible;

11) the following are constant and known:

- the production cost (C_{pr}),
- the inventory holding cost at S_1 (Cs_1),
- the inventory holding cost at S_0,
- the delivery cost (C_l),
- the preventive maintenance action cost (C_{pm}),
- the corrective maintenance action cost (C_{cm}).

Assumption 1 is usually supposed in the literature considering a stochastic demand distribution.

For assumption 2, it is normal to assume that the delivery time, τ, is multiple of Δt (the length of a period). For this reason we use a discrete-time model.

Assumption 3 is commonly made in manufacturing models that consider whether returned products (i.e. right of withdrawal) pass possible quality control or inspection conditions before the decision to restock is made.

Assumption 4 is often made in manufacturing models that consider returned products. Many works in the literature have proposed either normally distributed return rates or constant return rates.

For assumption 5, the return deadline is usually assumed to be constant in the literature.

Assumption 6 is also often made in manufacturing models that consider returned products. However, in our model the quantity of product returned by the customer is stochastic and the fact that it is proportional to demand is also stochastic.

Assumption 7 is common in production planning.

Assumption 8 is an original characteristic of our model, due to the consideration of a transport vehicle with limited capacity. This proposition makes our model more realistic.

In the literature it is usually assumed that after each preventive maintenance action, the equipment is in a state where it is "as good as new", as in assumption 9. We have assumed this because it allows us to analytically study the influence of production on both machine degradation and the average number of failures. This analytical study is difficult to resolve if we assume a policy without minimal repair. Maintenance is carried out at the end of the period as we assume that the delivery action is also applied at the end of the period.

Assumption 10 is proposed in order to simplify the analytical study of the maintenance. If we assume that the durations during which repair and overhaul are carried out are not negligible, we have to take into account lost production during maintenance procedures, and this makes the study very difficult.

Assumption 11 about constant and known unit costs is classical.

2.2.3. *Setting the problem*

In this section, a constrained stochastic optimization problem is formulated using a quadratic model by extending the work of Holt *et al.* [HOL 60]. The principle of the optimization method is to minimize the total expected cost over a finite time horizon $[0, H]$. In each period the demand is stochastic and follows Gaussian distribution. There is an average and a standard deviation given in each period. Demand is satisfied when the service level is met. This depends on the stock balance equation and remains stochastic due to demand.

As the service level constraint is stochastic, we can determine a minimum cumulative quantity that satisfies the customer. In what follows we present the stochastic problem, where $G_{USQ}(k)$ denotes the expected quadratic value of the sum of the inventory, delivery and production costs, while $C_{TM}(.)$ denotes the maintenance cost comprising both corrective and preventive maintenance costs.

$$Min(U,Q,N)\left(G_{USQ}(H) + \sum_{K=0}^{H-1} G_{USQ}(k) + C_{TM}(U,N) \right) \qquad [2.1]$$

REMARK 2.1.– we do not consider production control and transport at the end of the horizon, H; therefore $u(H) = Q(H) = 0$. Thus, we write the function $G_{USQ}(H)$ independently.

As mentioned before, we use a discrete time model to describe the system. At period k+1, therefore, the inventory balance level of S_1 is equal to the inventory level in the previous period (i.e. period k) plus the number of products produced by machine M in period k, minus the number of products transported from S_1 to S_0 in period k. Thus, the store level at period k+1 is described as follows:

$$S_1(k+1) = S_1(k) + u(k) - Q(k) \qquad [2.2]$$

with

$$k = \{0, 1, \ldots, H - 1\}$$

The inventory balance equation of S_0 is equation [2.3].

The quantity of parts entering the warehouse S_0 during period k is the quantity that left store S_1 in the period $k - \tau/\Delta t$. In other words, the transported parts that arrive at store S_0 in period k match the quantity of parts that left store S_1 in the period $k - \tau/\Delta t$ and is represented by $Q(k - \tau/\Delta t)$. Here, $\tau/\Delta t$ represents the number of periods that the products take to be transported between S_1 and S_0. However, to simplify the writing of the equations, we assume that $\Delta t = 1$, i.e. we have $Q(k - \tau)$ instead of $Q(k - \tau/\Delta t)$. Thus, during period $k+1$ the warehouse level is equal to the warehouse level during period k plus the parts entering S_0 having left store S_1, i.e. $Q(k - \tau)$, plus the parts the customer has returned, i.e. $r(k)$, minus the number parts requested by the customer in period k. Hence, the warehouse level in period $k+1$ is described as follows:

$$S_0(k+1)=\begin{cases} S_0(k)+Q(k-\tau)-d(k)+r(k) & \text{If } k\geq\tau \\ S_0(k)-d(k)+r(k) & \text{otherwise} \end{cases} \quad [2.3]$$

with

$k = \{0, 1,\ldots, H-1\}$ and $\tau \geq 1$.

REMARK 2.2.– we assume that $Q(k-\tau) = 0$ when $k<\tau$. When $k<\tau$, $k - \tau$ is negative (i.e. we have negative instant), this is impossible.

The quantity of returned parts denoted by $r(k)$ is a portion of the demand that is withdrawn by the customer after a specific period denoted by τ_r. In this case, $r(k)$ is presented as follows:

$$r(k)=\begin{cases} \delta\cdot d\left(k-\dfrac{\tau_r}{\Delta t}\right)=\delta\cdot d(k-\tau_r) & \text{if } k\geq\tau_r \\ 0 & \text{otherwise} \end{cases} \quad [2.4]$$

with

$k = \{0, 1,\ldots, H-1\}$ and $\tau_r \geq 1$.

REMARK 2.3.– in the case when $k<\tau_r$, the index of $d(k - \tau_r)$ is negative (i.e. the instant is negative). This situation is impossible, therefore we assume that $d(k - \tau_r) = 0$ when $k<\tau_r$, or in other words that $r(k) = 0$ when $k<\tau_r$.

The service level constraint that corresponds to a period k is described by the following constraint:

$$\text{Prob}\,(S_0(k+1) \geq 0) \text{ with } k = \{0, 1,\ldots, H\} \quad [2.5]$$

In what follows we present the constraint that defines the lower and upper bounds of machine production for a period, k:

$$U_{\min} \leq u(k) \leq U_{\max} \quad [2.6]$$

2.2.4. *Mathematical analysis*

In this section, we formulate an optimized constrained stochastic problem. We use this formulation to describe a constrained production–maintenance–delivery problem that takes account of:

– delivery time;

– product return;

– stochastic demand;

– rate of machine failure; and

– service level.

Inspired by the Modigliani–Muth–Simon model, we adapted the approach to formulate a stochastic model of this problem.

The model proposed by Holt *et al.* [HOL 60] presents a decision rule that determines an aggregate production, inventory and workforce policy. The proposed rule is determined by minimizing the quadratic production cost and describes the balance between production, inventory and workforce components. In our study we extend a quadratic model based on Holt *et al.* [HOL 60] in order to provide production, maintenance, inventory and delivery policies.

We now turn our attention to formulating expressions for the quadratic cost functions introduced in section 2.2.3.

2.2.4.1. *Derivation of expressions of production, inventory and delivery costs*

The main principle of the model is to define a quadratic cost function that penalizes both shortage and excess inventory. Our quadratic function is used to take into account the variation of a parameter of interest: inventory levels that are either too high or too low. It does this by using squared mathematical expectation.

In this chapter we have made some changes to the model, keeping its linear quadratic form. We take into account some constraints on the decision variables to make our approach more realistic and to ensure its applicability in real industrial cases.

Let the expected quadratic holding, production and delivery costs for period k be given by:

$$G_{USQ}(k) = G_U(k) + G_S(k) + G_Q(k) \quad [2.7]$$

where $E[\ ]$ denotes the mathematical expectation operator, and:

– the expected quadratic production cost for the number of products $u(k)$ is:

$$G_U(k) = C_{pr} \times E\left[u(k)^2\right] \quad [2.8]$$

– the expected quadratic holding cost for the inventory level in a period k is:

$$G_S(k) = C_{S1} \times E\left[S_1(k)^2\right] + C_{S0} \times E\left[S_0(k)^2\right] \quad [2.9]$$

– the expected quadratic delivery cost for the number of products transported, $Q(k)$, in a period k is:

$$G_Q(k) = C_l \times E\left[\left(\frac{Q(k)}{Q_v}\right)^2\right] \quad [2.10]$$

Using the quadratic approach, the expected total cost of inventory, delivery and production over the finite horizon *H*.Δt, denoted by GT_{USQ}, is described as follows:

$$GT_{USQ} = \sum_{k=0}^{H} G_{USQ}(k) =$$

$$C_{S1} \times E\left[S_1(H)^2\right] + C_{S0} \times E\left[S_0(H)^2\right] + \sum_{k=0}^{H-1} \left(\begin{array}{l} C_{pr} \times E\left[u(k)^2 \theta\right] + C_l \times E\left[\left(\dfrac{Q(k)}{Q_v}\right)^2\right] \\ + C_{S1} \times E\left[S_1(k)^2\right] + C_{S0} \times E\left[S_0(k)^2\right] \end{array} \right) \quad [2.11]$$

2.2.4.2. *Derivation of the expression of maintenance cost*

In this section we consider minimal repair by way of preventive maintenance over a finite horizon $H.\Delta t$. At times $h.T$ (with h = {0, 1,..., N}) we overhaul unit production, see Figure 2.2. Hence, the principle of the proposed overhaul is to replace several critical components in order to restore the production unit to a condition where it is "as-good-as-new". In other words, a minimal repair is carried out when the machine breaks down between successive overhauls.

There is a correlation between the delivery and production plans and system degradation: there is an increase in the rate of failure with an increase in the number of products produced by the machine and/or with longer periods of time [HAJ 09]. For this approach we elaborate on an analytical resolution that deals with the deterioration of the machine and does not only depend on the time but also on the rate of production. There is a correlation between the influence of variation in production and time on the degradation of the machine and on the average number of failures. Therefore, the advantage of this maintenance strategy is to improve machine availability and minimize maintenance costs in order to ensure production needs are met across horizon *H*. Other works develop the theory of the hedging point with a production rate that can only vary between 0 and the demand or the maximum production rate. In our work, we determine the rates by period to meet planning commitments while correlating the degradation of machine production rates. As these production rates vary from one period to another, the maintenance plan also changes to reflect the fact that degradation is not uniform over the planning horizon. The periodicities of preventive maintenance actions are thus recalculated to account for this.

A sequential optimization strategy is adopted to consider the maintenance strategy in order to obtain an optimal strategy that is characterized by the optimal number, N^*, of preventive maintenance actions required over the finite horizon, $H.\Delta t$.

The maintenance cost is represented by the following function:

$$C_{MT}\left(U=\{u(1),u(2),....u(H-1)\},N\right)=C_{pm}\times(N-1)+C_{cm}\times\varphi_M(U,N) \quad [2.12]$$

with $\varphi_M(U,N)$ being the average number of failures over the finite horizon, H.

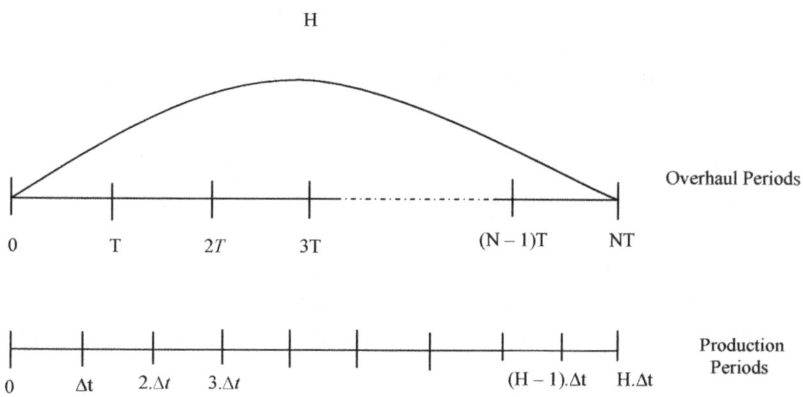

Figure 2.2. *Maintenance and production periods*

2.3. Transformation of the stochastic production, maintenance and delivery planning model to a deterministic equivalent

In this section, we transform the formulation from the previous section into a deterministic equivalent problem so that we may optimize it in section 2.4. Before presenting the transformation, however, we need to explain why we have decided to deal with this transformation.

2.3.1. *Motivation*

Many works deal with the transformation of a stochastic problem into deterministic equivalent in order to arrive at a solution. For example, through the steps of this transformation from stochastic problem into an equivalent deterministic we transform the probabilistic service level constraint into a deterministic constraint by specifying a minimum cumulative production quantity, depending on service level requirements.

The new part of our transformation deals with the transformation of a stochastic production, maintenance and delivery planning model into a deterministic equivalent. To the best of our knowledge this model, which establishes a production, maintenance and delivery planning problem, has never been dealt with in the literature.

2.3.2. *Transforming the production, inventory and delivery cost (expression [2.11]) into a deterministic equivalent*

In this section, we present the approach that we used to transform the stochastic problem into a deterministic equivalent. The deterministic problem retains the principal properties of the original problem.

The production, inventory and delivery quadratic costs can be simplified as demonstrated in Lemma 2.1.

LEMMA 2.1.–

$$\begin{aligned} GT_{USQ} = {} & \sum_{k=0}^{H}\left(C_{S1}\times\hat{S}_1(k)^2 + C_{S0}\times\hat{S}_0(k)^2\right) + C_{S0}\times\sum_{k=1}^{H}\left(\sigma_k^2\right)\times\left[H+1+(H-1)\times\delta^2\right]\times\frac{H}{2} \\ & + \sum_{k=0}^{H-1}\left(C_{pr}\times u(k)^2 + C_l\times\left(\frac{Q(k)}{Q_v}\right)^2\right) \end{aligned} \qquad [2.13]$$

The inventory balance equation [2.2] can be described as:

$$\hat{S}_1(k+1)=\hat{S}_1(k)+u(k)-Q(k)$$

with

$$k = \{0, 1,\ldots, H-1\}$$

and the inventory balance equation [2.3] can be described as:

$$\hat{S}_0(k+1)=\begin{cases}\hat{S}_0(k)+Q(k-\tau)+\delta\times\hat{d}(k-\tau_r)-\hat{d}(k) & \textit{If } k\geq\tau_r \textit{ and } k\geq\tau\\ \hat{S}_0(k)+\delta\times\hat{d}(k-\tau_r)-\hat{d}(k) & \textit{If } k\geq\tau_r \textit{ and } k<\tau\\ \hat{S}_0(k)+Q(k-\tau)-\hat{d}(k) & \textit{If } k<\tau_r \textit{ and } k\geq\tau\\ \hat{S}_0(k)-\hat{d}(k) & \textit{If } k<\tau_r \textit{ and } k<\tau\end{cases}$$

with

$$k = \{0, 1,\ldots, H-1\}, \text{ and } \tau_r\geq 1.$$

PROOF.– See Appendix 2.

2.3.3. *Transforming the service level constraint (equation [2.5]) into a deterministic equivalent*

In this section, we present the transformation of the stochastic service level constraint into its deterministic equivalent. Through the following lemma, we specify a minimum cumulative transportation quantity that depends on the service level.

LEMMA 2.2.–

This lemma is inspired by the Modigliani–Muth–Simon model used in by Holt *et al.* [HOL 60] and Silva Filho *et al.* [SIL 05].

$$\text{Prob}\left(S_0(k+1)\geq 0\right)\geq\theta \Rightarrow \left(Q(k-\tau)\geq\left(V_k\times V_{k-\tau_r}\right)\times\varphi^{-1}(\theta)-S_0(k)+\hat{d}(k)-\delta\times\hat{d}(k-\tau_r)\right)$$

with

$$k = \{0, 1, \ldots, H-1\} \qquad [2.14]$$

where φ is a cumulative Gaussian distribution function mean $\frac{1}{V_{k-\tau_r}}.\hat{d}(k)-\frac{\delta}{V_k}\cdot\hat{d}(k-\tau)$ and finite variance $\left(\left(\frac{1}{V_{k-\tau_r}}\right)^2 \times V_k + \left(-\frac{\delta}{V_k}\right)^2 \times V_{k-\tau_r} \geq 0\right)$, and φ^{-1} denotes the inverse distribution function.

PROOF.– According to equation [2.5], we have:

$$\text{Prob}(S_0(k+1) \geq 0))$$

In this case, the service level requirement constraint is given by:

$$\Rightarrow \text{Prob}\left(S_0(k)+Q(k-\tau)+\delta\times d(k-\tau_r)-d(k)\geq 0\right)\geq\theta$$

$$\Rightarrow \text{Prob}\left(S_0(k)+Q(k-\tau)\geq d(k)-\delta\times d(k-\tau_r)\right)\geq\theta$$

We add $\delta\times\hat{d}(k-\tau_r)$ to the expression to get:

$$\Rightarrow \text{Prob}\left(S_0(k)+Q(k-\tau)+\delta\times\hat{d}(k-\tau_r)-\hat{d}(k)\geq d(k)-\hat{d}(k)-\delta\times\left(d(k-\tau_r)-\hat{d}(k-\tau_r)\right)\right)\geq\theta$$

We divide the expression by $V_k \times V_{k-\tau_r}$ and then we have:

$$\Rightarrow \text{Prob}\left(\frac{S_0(k)+Q(k-\tau)+\delta\times\hat{d}(k-\tau_r)-\hat{d}(k)}{V_k\times V_{k-\tau_r}}\geq\frac{d(k)-\hat{d}(k)-\delta\times\left(d(k-\tau_r)-\hat{d}(k-\tau_r)\right)}{V_k\times V_{k-\tau_r}}\right)\geq\theta$$

Then

$$\Rightarrow \text{Prob}\left(\frac{1}{V_{k-\tau_r}}\times\frac{d(k)-\hat{d}(k)}{V_k}-\frac{\delta}{V_k}\times\frac{d(k-\tau_r)-\hat{d}(k-\tau_r)}{V_{k-\tau_r}}\leq\frac{S_0(k)+Q(k-\tau)+\delta\times\hat{d}(k-\tau_r)-\hat{d}(k)}{V_k\times V_{k-\tau_r}}\right)\geq\theta \qquad [2.15]$$

Note that:

– $X = \left(\frac{d(k) - \hat{d}(k)}{V_k} \right)$ is a Gaussian variable with a distribution identical to *d(k)*;

– $Y = \left(\frac{d(k-\tau_r) - \hat{d}(k-\tau_r)}{V_{k-\tau_r}} \right)$ is a Gaussian variable with an identical distribution to $d(k - \tau_r)$;

The formulation is then equivalent to:

$$\text{Prob}(A \times X + B \times Y \leq C) \geq \theta$$

$A = \frac{1}{V_{k-\tau_r}}$ and $B = -\frac{\delta}{V_k}$.

Note that $X' = A \times X$ is a Gaussian stochastic variable with an identical distribution to $f_{x'} = \frac{1}{A} \times f\left(\frac{y}{A}\right)$, with mean $A \times \hat{d}(k) = \frac{1}{V_{k-\tau}} \times \hat{d}(k)$ and variance $A^2 \times V_k = \left(\frac{1}{V_{k-\tau}}\right)^2 \times V_k \geq 0$, while $Y' = B \times Y$ is a Gaussian stochastic variable with an identical distribution to $f_{Y'} = -\frac{1}{B} \times f\left(\frac{y}{B}\right)$, with mean $B \times \hat{d}(k-\tau_r) = -\frac{\delta}{V_k} \times \hat{d}(k-\tau_r)$ and variance $B^2 \times V_{k-\tau_r} = \left(\frac{\delta}{V_k}\right)^2 \times V_{k-\tau_r} \geq 0$.

Thus we get a Gaussian stochastic variable with an identical distribution to $h = f_{X'} \times f_{Y'}$, $B \times \hat{d}(k-\tau_r) = -\frac{\delta}{V_k} \times \hat{d}(k-\tau_r)$ and variance $B^2 \times V_{k-\tau_r} = \left(\frac{\delta}{V_k}\right)^2 \times V_{k-\tau_r} \geq 0$.

The cumulative Gaussian distribution function of is denoted by φ.

Equation [2.15] implies that:

$$\varphi\left(\frac{S_0(k)+Q(k-\tau)+\delta\times\hat{d}(k-\tau_r)-\hat{d}(k)}{V_k\times V_{k-\tau_r}}\right)\geq\theta \qquad [2.16]$$

and we conclude that it is strictly increasing. We note that the equation is indefinitely differentiable, so we conclude that it is invertible.

Equation [2.16] implies that:

$$\frac{S_0(k)+Q(k-\tau)+\delta\times\hat{d}(k-\tau_r)-\hat{d}(k)}{V_k\times V_{k-\tau_r}}\geq\varphi^{-1}\times(\theta)$$

therefore:

$$S_0(k)+Q(k-\tau)+\delta\times\hat{d}(k-\tau_r)-\hat{d}(k)\geq V_k\times V_{k-\tau_r}\times\varphi^{-1}\times(\theta)$$

$$Q(k-\tau)\geq V_k\times V_{k-\tau_r}\times\varphi^{-1}(\theta)-S_0(k)-\delta\times\hat{d}(k-\tau_r)+\hat{d}(k)$$

thus:

$$\text{Prob}\left(S_0(k+1)\geq 0\right)\geq\theta\ \Rightarrow\left(Q(k-\tau)\geq V_k\times V_{k-\tau_r}\times\varphi^{-1}\times(\theta)-S_0(k)-\delta\times\hat{d}(k-\tau_r)+\hat{d}(k)\right)$$

Q.E.D.

2.3.4. *Transforming the maintenance cost (expression [2.12]) into a deterministic equivalent*

In this section, the optimal maintenance strategy is determined. This means that we will determine the optimum number of preventive maintenance actions (N^*).

$$T^*\times N^*=H \qquad [2.17]$$

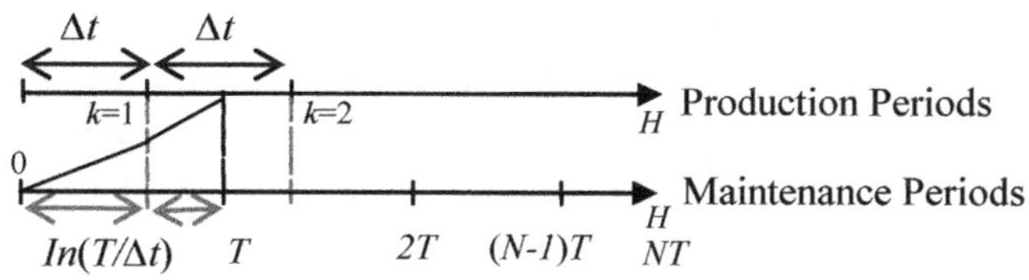

Figure 2.3. *Evolution of the rate of failure over time. For a color version of this figure, see www.iste.co.uk/rezg/services.zip*

The analytical expression of the total maintenance cost is described as follows:

$$C_{TM}(U,N) = (N-1) \times C_{pm} + C_{cm} \times \varphi_M(U,N) \qquad [2.18]$$

where $N \in \{1,2,3.....\}$, and $\varphi_M(U,N)$ corresponds to the expected number of failures that occur in the horizon H with consideration of the number of products produced by machine M during every production period Δt.

Recall that $\lambda_k(t)$ represents the function of the linear failure rate at production period k (see Figure 2.3), while $\lambda_n(t)$ is the nominal condition failure rate, which is known as the maximal production failure rate. Then we have:

$$\lambda_k(t) = \lambda_{k-1}(\Delta t) + \frac{u(k)}{U_{\max}} \times \lambda_n(t) \qquad [2.19]$$

This equation shows the evolution of the machine's failure rate according to use (which in our case is the production rate for each period) respecting the continuity of equipment reliability from one period to another. This equation has been developed in Hajej *et al.* [HAJ 14b]. From this equation, we have determined the average number of failures, assuming that after each preventive maintenance action the equipment is "as good as new" and that maintenance action may be applied during production as well as at the end of the period.

In what follows, we present the average number of failures over the finite horizon, which is a function of the failure rate.

Let *In* denote the integer part of (.). The average number of failures over the horizon, *H*, is:

$$\varphi_M(U,T)=\sum_{j=0}^{N-1}\left[\sum_{i=In(j\times\frac{T}{\Delta t})+1}^{In\left((j+1)\times\frac{T}{\Delta t}\right)}\int_0^{\Delta t}\lambda_i(t)+\int_0^{(j+1)\times T-In\left((j+1)\times\frac{T}{\Delta t}\right)\times\Delta t}\lambda_{In\left((j+1)\times\frac{T}{\Delta t}\right)+1}(t)\,dt\right.$$

$$\left.+\int_{(j+1)\times T}^{\left(In\left((j+1)\times\frac{T}{\Delta t}\right)+1\right)\times\Delta t}\frac{\left(In\left((j+1)\times\frac{T}{\Delta t}\right)+1\right)}{U_{\max}}\times\lambda_n(t)\,dt\right]\qquad [2.20]$$

and

$$\lambda_i(t)=\lambda_{i-1}(\Delta t)+\frac{u(i)}{U_{\max}}\times\lambda_n(t)\qquad \forall t\in[0,\Delta t]$$

$$\Rightarrow\lambda_i(t)=\lambda_0(t_0)+B_i+\frac{u(i)}{U_{\max}}\times\lambda_n(t)$$

$$B_i=\sum_{l=1}^{i-1}\frac{u(l)}{U_{\max}}\times\lambda_n(\Delta t)$$

where $B_1=0$ and $\lambda_0(t_0)=\lambda_0$.

Therefore, we replace $\lambda_i(t)$ in expression [2.20]:

$$\varphi_M(U,T)=\sum_{j=0}^{N-1}\left(\begin{array}{l}\left(In\left((j+1)\times\frac{T}{\Delta t}\right)-In\left(j\times\frac{T}{\Delta t}\right)\right)\times\Delta t\times\lambda_0(t_0)+\\[2ex]\sum_{i=In\left(j\times\frac{T}{\Delta t}\right)+1}^{In\left((j+1)\times\frac{T}{\Delta t}\right)}\int_0^{\Delta t}\left(\sum_{l=1}^{i-1}\frac{u(l)}{U_{max}}\times\lambda_0(\Delta t)\right)dt+\sum_{i=In\left(j\times\frac{T}{\Delta t}\right)+1}^{In\left((j+1)\times\frac{T}{\Delta t}\right)}\int_0^{\Delta t}\frac{u(i)}{U_{max}}\times\lambda_0(t)\,dt\\[2ex]+\int_0^{(j+1)\times T-In\left(\frac{(j+1)\times T}{\Delta t}\right)\times\Delta t}\left(\sum_{l=1}^{In\left(\frac{(j+1)\times T}{\Delta t}\right)}\frac{u(l)}{U_{max}}\times\lambda_0(\Delta t)\right)dt\\[2ex]+\int_0^{(j+1)\times T-In\left(\frac{(j+1)\times T}{\Delta t}\right)\times\Delta t}\frac{u\left(In\left(\frac{(j+1)\times T}{\Delta t}\right)+1\right)}{U_{max}}\times\lambda_0(t)\,dt\\[2ex]+\frac{u\left(In\left(\frac{(j+1)\times T}{\Delta t}\right)+1\right)}{U_{max}}\times\int_{In\left((j+1)\times\frac{T}{\Delta t}\right)\times\Delta t}^{(j+1)T}\lambda_0(t)\,dt\end{array}\right)$$

We now replace $T = H/N$:

$$\varphi_M\left(U,H/N\right)=\sum_{j=0}^{N-1}\left(\begin{array}{l}\left(In\left((j+1)\times\frac{H}{N\times\Delta t}\right)-In\left(j\times\frac{H}{N\times\Delta t}\right)\right)\times\Delta t\times\lambda_0(t_0)+\\[2ex]\frac{\lambda_0(\Delta t)\times\Delta t}{U_{max}}\times\sum_{i=In\left(j\times\frac{T}{\Delta t}\right)+1}^{In\left((j+1)\times\frac{H}{N\times\Delta t}\right)}\sum_{l=1}^{i-1}u(l)\,dt+\\[2ex]\frac{1}{U_{max}}\times\sum_{i=In\left(j\times\frac{H}{N\times\Delta t}\right)+1}^{In\left((j+1)\times\frac{H}{N\times\Delta t}\right)}\int_0^{\Delta t}u(i)\times\lambda_0(t)\,dt+\\[2ex]\sum_{l=1}^{In\left(\frac{(j+1)\times H}{N\times\Delta t}\right)}\frac{u(l)}{U_{max}}\times\lambda_0(\Delta t)\times\left((j+1)\times\frac{H}{N}-In\left(\frac{(j+1)\times H}{N\times\Delta t}\right)\times\Delta t\right)\\[2ex]+\int_0^{(j+1)\times\frac{H}{N}-In\left(\frac{(j+1)\times H}{N\times\Delta t}\right)\times\Delta t}\frac{u\left(In\left(\frac{(j+1)\times H}{N\times\Delta t}\right)+1\right)}{U_{max}}\times\lambda_0(t)\,dt\\[2ex]+\frac{u\left(In\left(\frac{(j+1)\times H}{N\times\Delta t}\right)+1\right)}{U_{max}}\times\int_{In\left((j+1)\times\frac{H}{N\times\Delta t}\right)\times\Delta t}^{(j+1)\times\frac{H}{N}}\lambda_0(t)\,dt\end{array}\right)$$

In the following section we are interested in studying the impact of delivery time, the rate of product return and service level on the optimal production and delivery plans and on the optimal maintenance strategy. The model we have derived is used for the optimization.

2.4. Numerical example and numerical optimization procedure

In this section, some numerical examples are presented to study the impact of different model parameters on optimal production and delivery quantity plans by varying parameters such as delivery time, rate of returned products and service level. Then, the results are interpreted and discussed. Firstly, the principle of used optimization numerical procedure is presented followed by different numerical examples concerning our model.

2.4.1. *Numerical optimization procedure*

Using the stochastic model formulation presented in the previous sections, we formulate a deterministic equivalent problem for the integrated optimization problem of inventory, production, maintenance and delivery quadratic costs as follows:

$$\underset{(U,Q,N)}{Min}\left(\begin{array}{l}\sum_{k=0}^{H}\left(C_{s1}\cdot\hat{S}_1(k)^2+C_{s0}\cdot\hat{S}_0(k)^2\right)+\\ \sum_{k=1}^{H-1}\left(C_{p_r}\times u(k)^2+C_l\times\left(\dfrac{Q(k)}{Q_v}\right)^2\right)\\ +C_{s0}\cdot\sigma_d^2\cdot\left[H+1+(H-1)\cdot\delta^2\right]\cdot\dfrac{H}{2}\\ +(N-1)\cdot C_{pm}+C_{cm}\cdot\varphi_M(U,N)\end{array}\right) \quad [2.21]$$

Subject to:

$$\hat{S}_1(k+1)=\hat{S}_1(k)+u(k)-Q(k) \text{ with } k=\{0,1,\ldots,H\text{-}1\} \quad [2.22]$$

$$\hat{S}_0(k+1)=\hat{S}_0(k)+Q(k-\tau)+\delta\times\hat{d}(k-\tau_r)-\hat{d}(k) \text{ with } k=\{0, 1, \ldots, H\text{-}1\} \quad [2.23]$$

$$Q(k-\tau)\geq\left(V_k\times V_{k-\tau_r}\right)\times\varphi^{-1}(\theta)-S_0(k)+\hat{d}(k)-\delta\times\hat{d}(k-\tau_r) \text{ with } k=\{0, 1, \ldots, H\text{-}1\} \quad [2.24]$$

$$U_{\min}\leq u(k)\leq U_{\max} \text{ with } k=\{0, 1, \ldots, H\text{-}1\} \quad [2.25]$$

Due to the nonlinearity of the problem constraints and in order to realize this optimization, we used a Numerical Algorithms for Constrained Global Optimization based on the Nelder Mead methods. Indeed, The Nelder–Mead technique was proposed by John Nelder and Roger Mead [NED 65] and is a nonlinear optimization technique for minimising an objective function in a many-dimensional space and which used for solving problems that derivatives may not be known. Furthmore, the Nelder–Mead technique is a heuristic search method that can converge to non-stationary points on problems that can be solved by alternative methods. Lagarias *et al.* [LAG 98] presented the convergence properties of the Nelder-Mead algorithm applied to functions in dimensions 1 and 2. The authors proved the convergence to a minimiser for dimension 1, and various limited convergence results for dimension 2. Fazekas *et al.* [FAZ 08] defined the principle of the Nelder-Mead algorithm as the following. A simplex is the convex hull of $n+1$ vertex in an n-dimensional space. The method starts from an initial working simplex which is created using the given initial parameter value. The algorithm then performs a sequence of transformations (that can be reflection, expansion, contraction or shrink) of the working simplex, to decrease the objective function values at the vertices. The algorithm is terminated when the size of the simplex is sufficiently small, or when the function values at the vertices are close to each other in some norm. In each iteration step, the algorithm typically needs only one or two objective function evaluations which are quite low compared to most other methods. For more details about the Nelder-Mead method, we cite the article by Yang *et al.* [YAN 12] which presented an algorithm that describes the general Nelder-Mead simplex method. The simplex search algorithm does not guarantee that the obtained point is a global minimum on the whole parameter domain (as many nonlinear optimization techniques).

Thus, it is important to use as much prior information about the modeled process as possible to choose proper initial parameter values for the method.

The following numerical optimization procedure is proposed to solve the problem. It has five steps:

– Step 1: Introduce input data (demand, production cost, inventory cost, delivery cost, corrective maintenance cost, preventive maintenance cost, etc.)

– Step 2: Use equation [2.24], to determine the minimum cumulative transportation quantity that depends on the service level constraint given by equation [2.25].

– Step 3: Choose a production vector $U = \{u(1),u(2),\ldots,u(H\text{-}1)\}$, $Q = \{Q(1),Q(2),\ldots,Q(H\text{-}1)\}$ by varying the production quantity and the delivery quantity for each period according to constraints [2.24] and [2.25] and the number of preventive maintenance actions N in order to consider all possible production, delivery and maintenance plans.

– Step 4: For every plan established in step 3, we calculate each time the corresponding total cost (equation [2.21]).

– Step 5: Determine the optimal values of the decision variables which yielded the minimal total cost.

2.4.2. *Numerical example*

The different input data are summarized in Table 2.1 below ("mu" stands for monetary units):

Reliability data	
System time to failure distribution	F(t): Weibull distribution with : Scale parameter : $\beta = 100$ & Shape parameter : $\gamma = 2$
Corrective maintenance cost	$Ccm = 3000$ *mu.*
Preventive maintenance cost	$C_{pm} = 500$ *mu*
Inventory data	
Inventory holding cost for S_1	$Cs_1 = 0.2$ mu.
Inventory holding cost for S_0	$Cs_0 = 0.2$ mu.

Initial inventory level for S_0	$S_0(0) = 20$
Initial inventory level for S_1	$S_1(0) = 50$
Service level	$\theta = 0.9$
Production data	
Finite horizon of production	$H = 36$ months
Length of production period	$\Delta t = 1$ *months*
Unit production cost	$Cpr = 2$ *mu.*
Minimal production rate	$Umin = 5.$
Maximal production rate	$U_{\max} = 17.$
Delivery data	
Delivery cost	$C_l = 12$ *mu*
Delivery vehicle capacity	$Q_v = 6$
Time between the customer receiving and products returning	$\tau_r = 1$
Rate of returned product	$\delta = 0.3$

Table 2.1. *Numerical data*

Example 1

In this first example, we assume that the demand is generated by Gaussian distribution with standard deviation $\sigma_d = 1.2$, and the average demand for each period Δt is given by the following Table 2.2.

$\hat{d}(0)$	$\hat{d}(1)$	$\hat{d}(2)$	$\hat{d}(3)$	$\hat{d}(4)$	$\hat{d}(5)$
15	**17**	**15**	**15**	**15**	**14**
$\hat{d}(6)$	$\hat{d}(7)$	$\hat{d}(8)$	$\hat{d}(9)$	$\hat{d}(10)$	$\hat{d}(11)$
16	**14**	**16**	**15**	**15**	**15**
$\hat{d}(12)$	$\hat{d}(13)$	$\hat{d}(14)$	$\hat{d}(15)$	$\hat{d}(16)$	$\hat{d}(17)$
15	**15**	**15**	**13**	**15**	**15**
$\hat{d}(18)$	$\hat{d}(19)$	$\hat{d}(20)$	$\hat{d}(21)$	$\hat{d}(22)$	$\hat{d}(23)$
16	**13**	**15**	**15**	**14**	**16**
$\hat{d}(24)$	$\hat{d}(25)$	$\hat{d}(26)$	$\hat{d}(27)$	$\hat{d}(28)$	$\hat{d}(29)$
16	**16**	**14**	**15**	**15**	**14**
$\hat{d}(30)$	$\hat{d}(31)$	$\hat{d}(32)$	$\hat{d}(33)$	$\hat{d}(34)$	$\hat{d}(35)$
15	**16**	**14**	**16**	**14**	**14**

Table 2.2. *Average demand for the numerical example*

Using a numerical procedure based on the Nelder-Mead algorithm programmed on MATHEMATICA® software which has been used to perform the calculations and obtain the optimal solution for any given instance of the problem. However, in our case and by using the Nelder-Mead algorithm, we just enter the values set of each parameter (i.e. the possible parameter values vector). Thus, in this numerical example, the values sets are : for the production we have $U_{min} = 5 \leq u \leq U_{max} = 17$ (i.e. $u = \{5,6...,17\}$), for the transported quantity we have $0 \leq Q \leq Qv = 6$ (i.e. $u = \{0,1...,6\}$) and the maintenance we have $1 \leq N \leq H = 36$ (i.e. $N = \{1,2...,36\}$). Then, the numerical algorithm based on the Nelder-Mead methods chooses randomly the initial parameter values and run the simplex method repeatedly with different initial parameter values to find a local extremum. The minimum among resultant local extreme values is selected as a solution. Therefore, this algorithm finds the minimum of the objective function while guaranteeing the optimality of the solutions.

The optimal production, delivery and maintenance plans are presented respectively in Tables 2.3 and 2.4 as well as in Figure 2.4.

$u^*(0)$	$u^*(1)$	$u^*(2)$	$u^*(3)$	$u^*(4)$	$u^*(5)$
8	**12**	**17**	**17**	**17**	**12**
$u^*(6)$	$u^*(7)$	$u^*(8)$	$u^*(9)$	$u^*(10)$	$u^*(11)$
12	**10**	**7**	**13**	**17**	**11**
$u^*(12)$	$u^*(13)$	$u^*(14)$	$u^*(15)$	$u^*(16)$	$u^*(17)$
16	**16**	**13**	**12**	**11**	**9**
$u^*(18)$	$u^*(19)$	$u^*(20)$	$u^*(21)$	$u^*(22)$	$u^*(23)$
5	**5**	**11**	**10**	**17**	**5**
$u^*(24)$	$u^*(25)$	$u^*(26)$	$u^*(27)$	$u^*(28)$	$u^*(29)$
13	**17**	**17**	**14**	**17**	**11**
$u^*(30)$	$u^*(31)$	$u^*(32)$	$u^*(33)$	$u^*(34)$	$u^*(35)$
7	**7**	**16**	**9**	**8**	**6**

Table 2.3. *Optimal production plan*

$Q^*(0)$	$Q^*(1)$	$Q^*(2)$	$Q^*(3)$	$Q^*(4)$	$Q^*(5)$
30	**25**	**5**	7	**17**	**5**
$Q^*(6)$	$Q^*(7)$	$Q^*(8)$	$Q^*(9)$	$Q^*(10)$	$Q^*(11)$
5	**16**	**3**	**27**	**0**	**21**
$Q^*(12)$	$Q^*(13)$	$Q^*(14)$	$Q^*(15)$	$Q^*(16)$	$Q^*(17)$
18	**10**	**6**	**13**	**0**	**11**
$Q^*(18)$	$Q^*(19)$	$Q^*(20)$	$Q^*(21)$	$Q^*(22)$	$Q^*(23)$
10	**0**	**8**	**29**	**0**	**5**
$Q^*(24)$	$Q^*(25)$	$Q^*(26)$	$Q^*(27)$	$Q^*(28)$	$Q^*(29)$
13	**20**	**10**	**10**	**15**	**27**
$Q^*(30)$	$Q^*(31)$	$Q^*(32)$	$Q^*(33)$	$Q^*(34)$	$Q^*(35)$
11	**8**	**1**	**0**	**0**	7

Table 2.4. *Optimal delivery plan*

Tables 2.3 and 2.4 illustrate the optimal production and delivery plans that minimize the total cost of the system in order to satisfy the stochastic demand under the given service level. As we see in the Table 2.3, the production quantity of course changes from one period to another and this is due to the variability of the demand. We see that in the periods when $k = \{2, 3, 4, 10, 22, 25, 26, 28\}$, the production takes the maximum value and this is due to that our model forecasts the production quantity according to the demand value under the service level constraint. For the optimal delivery plan which is presented in the Table 2.4, is determined based on the relationship with the production/maintenance planning and the service level. Indeed, our model forecasts for each period the quantity to deliver between the manufacturing store and a purchase warehouse and this is according to the production quantity and the demand value.

We present in Figure 2.4 the total cost as a function of the delivery time τ and number of the preventive maintenance actions N. As we see that the lowest total cost value corresponds to $\tau^* = 2$ and $N^* = 2$. Therefore, over the finite horizon H of 36 months, two preventive maintenance actions should be done, i.e. at every T* = H/N^* = 18 months.

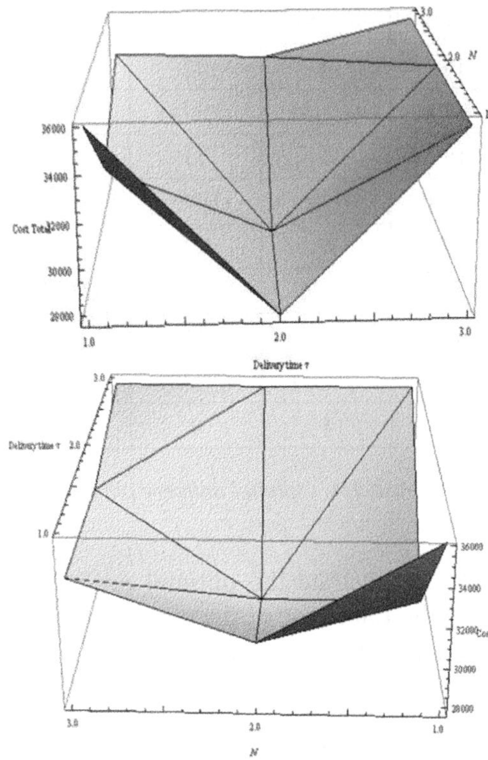

Total Cost*=27716.4 mu, N*=2,τ*=2

Figure 2.4. *The total cost as a function of the delivery time τ and number of preventive maintenance actions N*

Example 2

In this second example, using the same data as example 1, we will treat the case of a new production system composed by a machine that characterized by minimal production rate equals to $Umin$ = 100 and maximal production rate U_{max} = 500. The rates of new forecasting demands are higher than the first demands of example 1 and are characterized by a standard deviation equals to σ_d= 5 for each period and their averages are given by the following table 5. Also, we assume that initial inventory level for S_0 = 100 and for S_1 = 200.

$\hat{d}(0)$	$\hat{d}(1)$	$\hat{d}(2)$	$\hat{d}(3)$	$\hat{d}(4)$	$\hat{d}(5)$
350	420	340	392	431	444
$\hat{d}(6)$	$\hat{d}(7)$	$\hat{d}(8)$	$\hat{d}(9)$	$\hat{d}(10)$	$\hat{d}(11)$
442	340	392	375	392	400
$\hat{d}(12)$	$\hat{d}(13)$	$\hat{d}(14)$	$\hat{d}(15)$	$\hat{d}(16)$	$\hat{d}(17)$
350	370	395	415	431	444
$\hat{d}(18)$	$\hat{d}(19)$	$\hat{d}(20)$	$\hat{d}(21)$	$\hat{d}(22)$	$\hat{d}(23)$
442	340	392	375	400	420
$\hat{d}(24)$	$\hat{d}(25)$	$\hat{d}(26)$	$\hat{d}(27)$	$\hat{d}(28)$	$\hat{d}(29)$
350	340	392	370	431	392
$\hat{d}(30)$	$\hat{d}(31)$	$\hat{d}(32)$	$\hat{d}(33)$	$\hat{d}(34)$	$\hat{d}(35)$
500	350	320	420	365	480

Table 2.5. *Average demand*

The optimal production and delivery plans are presented in Tables 2.6 and 2.7 respectively. According the obtained economical production plan and the relationship of failure rate, we optimize the maintenance strategy by determining the optimal number of preventive maintenance actions which equals to $N^* = 3$ corresponding to the optimal delivery time $\tau^* = 2$ presented in Figure 2.5. Comparing by the first example, we can note that the optimal number of preventive maintenance and the total cost are higher. We can interpret that production rates increase then the failure rate increases as well as the average number of failure and consequently the number of preventive maintenance actions.

$u^*(0)$	$u^*(1)$	$u^*(2)$	$u^*(3)$	$u^*(4)$	$u^*(5)$
500	**341**	**417**	**406**	**100**	**370**
$u^*(6)$	$u^*(7)$	$u^*(8)$	$u^*(9)$	$u^*(10)$	$u^*(11)$
500	**361**	**399**	**175**	**457**	**500**
$u^*(12)$	$u^*(13)$	$u^*(14)$	$u^*(15)$	$u^*(16)$	$u^*(17)$
100	**468**	**100**	**500**	**447**	**145**
$u^*(18)$	$u^*(19)$	$u^*(20)$	$u^*(21)$	$u^*(22)$	$u^*(23)$
500	**107**	**230**	**372**	**100**	**255**
$u^*(24)$	$u^*(25)$	$u^*(26)$	$u^*(27)$	$u^*(28)$	$u^*(29)$
500	**100**	**339**	**445**	**114**	**100**
$u^*(30)$	$u^*(31)$	$u^*(32)$	$u^*(33)$	$u^*(34)$	$u^*(35)$
100	**347**	**277**	**397**	**500**	**226**

Table 2.6. *Optimal production plan*

$Q^*(0)$	$Q^*(1)$	$Q^*(2)$	$Q^*(3)$	$Q^*(4)$	$Q^*(5)$
525	**524**	**500**	**500**	**500**	**500**
$Q^*(6)$	$Q^*(7)$	$Q^*(8)$	$Q^*(9)$	$Q^*(10)$	$Q^*(11)$
576	**500**	**0**	**1000**	**245**	**111**
$Q^*(12)$	$Q^*(13)$	$Q^*(14)$	$Q^*(15)$	$Q^*(16)$	$Q^*(17)$
371	**0**	**658**	**0**	**328**	**296**
$Q^*(18)$	$Q^*(19)$	$Q^*(20)$	$Q^*(21)$	$Q^*(22)$	$Q^*(23)$
132	**0**	**496**	**0**	**266**	**0**
$Q^*(24)$	$Q^*(25)$	$Q^*(26)$	$Q^*(27)$	$Q^*(28)$	$Q^*(29)$
527	**539**	**441**	**0**	**514**	**400**
$Q^*(30)$	$Q^*(31)$	$Q^*(32)$	$Q^*(33)$	$Q^*(34)$	$Q^*(35)$
417	**500**	**695**	**0**	**88**	**0**

Table 2.7. *Optimal delivery plan*

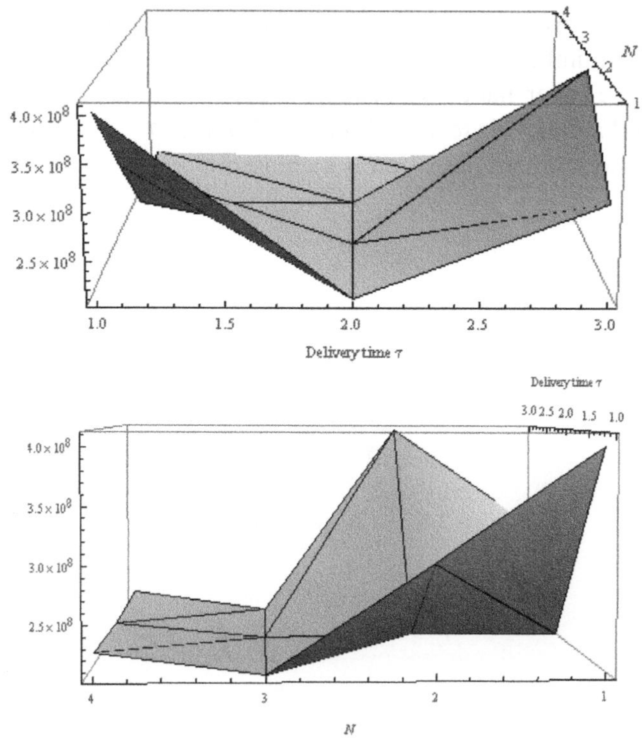

Total Cost* $= 2.05807 \times 10^8$ *mu*, ***N**** $= 3$, $\tau^* = 2$

Figure 2.5. *The total cost as a function of the delivery time τ and number of preventive maintenance actions N*

Example 3

Using the same data as previous example but by changing the value of the standard deviation of random demand, we determine the economical production and delivery quantity plans for each value of σ (σ = 2.5, σ = 5, σ = 10).

From Figures 2.6 and 2.7 and Table 2.8, we note that, as standard deviation σ increases, the economical plans for the production and delivery quantity increase as well as the total cost because the machine will produce more in order to satisfy the random demand under given service level and consequently the degradation degree of machine increases as well as the average number of failure and this need more number of preventive maintenance actions ($N^* = 4$ for $\sigma = 10$). On the other hand, we remark that the optimal delivery time is unchanged and equals to $\tau^* = 2$ for $\sigma = 2.5$ and $\sigma = 5$ and decreases to $\tau^* = 1$ if the standard deviation is increased $\sigma = 10$.

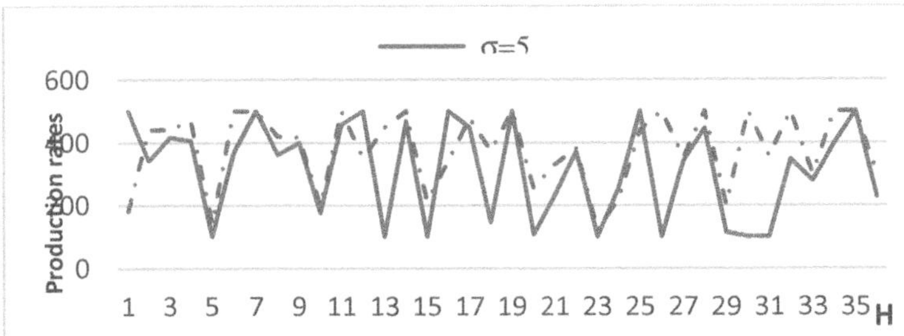

Figure 2.6. *Economical production plan*

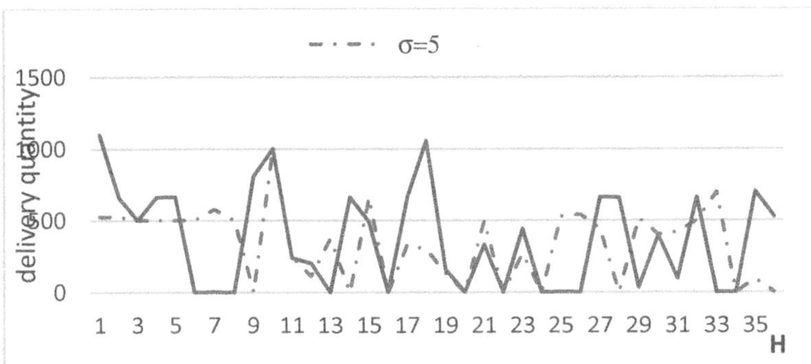

Figure 2.7. *Delivery quantity plan*

Standard deviation σ	Total Cost	Optimal number of preventive maintenance action N^*	Delivery time τ
2.5	2.20037×10^8	2	2
5	2.05807×10^8	3	2
10	4.00356×10^8	4	1

Table 2.8. *Delivery time τ and number of preventive maintenance actions N for each value of σ*

Example 4

This last example treats the impact of the variation of corrective maintenance cost unit *Ccm* on the maintenance strategy of the optimization problem. In this case, we determine the different optimization results, characterized by the optimal number of preventive maintenance actions *N**, by varying the value of *Ccm* (*Ccm = 2000 mu, Ccm = 3000 mu, Ccm = 6000 mu*) with unit preventive maintenance cost equals to *Cpm* = 500. Table 2.9 shows that the higher value of number of preventive maintenance actions (N* = 5) corresponds to the higher value of corrective maintenance cost (*Ccm* = 6000). This can be explained by the fact that, as the corrective maintenance cost increases, the optimal number of preventive maintenance actions increase in order to reduce the failure rate since the corrective maintenance cost become expensive.

Corrective maintenance cost unit Ccm	Total cost	Optimal number of preventive maintenance action N*
2000	2.00356 × 108	2
3000	2.05807 × 108	3
6000	4.00356 × 108	5

Table 2.9. *Number of preventive maintenance actions N for each value of Ccm*

2.4.3. *Variability study of delivery time, returned products and service level*

In this section, we study the impact of delivery time, rate of returned products and service level on the optimal production and optimal delivery quantity plans. However, by using the previous optimization procedure, we change the values of the delivery time, rate of returned products and service level and then we determine the corresponding optimal plans in order to study the variability of those parameters.

The following parameters are used for the simulations (Table 2.10).

Reliability data	
System time to failure distribution	F(t): Weibull distribution with: Scale parameter: $\beta = 100$ & Shape parameter: $\gamma = 2$
Corrective maintenance cost	$Ccm = 3500\ mu$
Preventive maintenance cost	$C_{pm} = 100\ mu$
Inventory data	
Inventory holding cost for S_1	$Cs_1 = 3\ mu$
Inventory holding cost for S_0	$Cs_0 = 3\ mu$
Initial inventory level for S_0	$S_0(0) = 20$
Initial inventory level for S_1	$S_1(0) = 50$
Production data	
Finite horizon of production	$H = 36$ months
Length of production period	$\Delta t = 1\ months$
Unit production cost	$C_{pr} = 2\ mu$
Minimal production rate	$U_{\min} = 0$
Maximal production rate	$U_{\max} = 22$
Delivery data	
Delivery cost	$Cl = 12\ m$
Delivery vehicle capacity	$Q_v = 6$
Time between the customer receiving and products returning	$\tau_r = 1$

Table 2.10. *Numerical data*

The service level $\theta \in [0.5, 1]$, rate of backordered product $\delta \in [0, 0.5]$ and returned production deadline $\tau_r = 1$ (for the simulation of the variability of returned products rate and service level).

In what follows, we are interested in finding the optimal production plan and the optimal delivery quantity plan for each value of delivery time τ in order to satisfy the random demand given in Table 2.2. Also, we are interested in finding the optimal delivery time which minimizes the total cost and then we will provide interpretation.

Example 1 (Variability of delivery time)

We used the Numerical Algorithms for Constrained Global Optimization (Nelder–Mead methods) with MATHEMATICA to run the problem optimization. The economical production plan and the variability study are presented respectively in Tables 2.11–2.16 and Figures 2.8–2.10.

Tables 2.11–2.16 illustrate example results for an optimal production plan and an optimal transported quantity which minimise the total cost for different values of the delivery time τ ($\tau = 1$, $\tau = 2$, $\tau = 3$).

Case 1: Delivery Time τ = 1

Cost = 149230mu / N* = 2

$u^*(0)$	$u^*(1)$	$u^*(2)$	$u^*(3)$	$u^*(4)$	$u^*(5)$
22	**22**	**22**	**8**	**1**	**17**
$u^*(6)$	$u^*(7)$	$u^*(8)$	$u^*(9)$	$u^*(10)$	$u^*(11)$
7	**3**	**10**	**17**	**11**	**6**
$u^*(12)$	$u^*(13)$	$u^*(14)$	$u^*(15)$	$u^*(16)$	$u^*(17)$
4	**2**	**0**	**14**	**13**	**22**
$u^*(18)$	$u^*(19)$	$u^*(20)$	$u^*(21)$	$u^*(22)$	$u^*(23)$
9	**3**	**10**	**22**	**9**	**16**
$u^*(24)$	$u^*(25)$	$u^*(26)$	$u^*(27)$	$u^*(28)$	$u^*(29)$
0	**17**	**16**	**0**	**4**	**9**
$u^*(30)$	$u^*(31)$	$u^*(32)$	$u^*(33)$	$u^*(34)$	-
22	**1**	**9**	**10**	**17**	-

Table 2.11. *Principal machine u*(k)*

$Q^*(0)$	$Q^*(1)$	$Q^*(2)$	$Q^*(3)$	$Q^*(4)$	$Q^*(5)$
25	**23**	**11**	**13**	**6**	**19**
$Q^*(6)$	$Q^*(7)$	$Q^*(8)$	$Q^*(9)$	$Q^*(10)$	$Q^*(11)$
0	**25**	**14**	**14**	**0**	**0**
$Q^*(12)$	$Q^*(13)$	$Q^*(14)$	$Q^*(15)$	$Q^*(16)$	$Q^*(17)$
16	**17**	**16**	**7**	**1**	**12**
$Q^*(18)$	$Q^*(19)$	$Q^*(20)$	$Q^*(21)$	$Q^*(22)$	$Q^*(23)$
17	**1**	**25**	**5**	**17**	**19**
$Q^*(24)$	$Q^*(25)$	$Q^*(26)$	$Q^*(27)$	$Q^*(28)$	$Q^*(29)$
2	**0**	**12**	**11**	**16**	**0**
$Q^*(30)$	$Q^*(31)$	$Q^*(32)$	$Q^*(33)$	$Q^*(34)$	–
15	**21**	**0**	**22**	**7**	–

Table 2.12. *Transported quantity: Q*(k)*

Case 2: Delivery Time $\tau = 2$

Cost* = 134987mu / N* = 2

$u^*(0)$	$u^*(1)$	$u^*(2)$	$u^*(3)$	$u^*(4)$	$u^*(5)$
18	**22**	**22**	**22**	**9**	**4**
$u^*(6)$	$u^*(7)$	$u^*(8)$	$u^*(9)$	$u^*(10)$	$u^*(11)$
17	**3**	**13**	**22**	**11**	**19**
$u^*(12)$	$u^*(13)$	$u^*(14)$	$u^*(15)$	$u^*(16)$	$u^*(17)$
9	**0**	**9**	**0**	**17**	**7**
$u^*(18)$	$u^*(19)$	$u^*(20)$	$u^*(21)$	$u^*(22)$	$u^*(23)$
3	**20**	**10**	**22**	**11**	**0**
$u^*(24)$	$u^*(25)$	$u^*(26)$	$u^*(27)$	$u^*(28)$	$u^*(29)$
22	**5**	**3**	**10**	**12**	**12**
$u^*(30)$	$u^*(31)$	$u^*(32)$	$u^*(33)$	–	–
8	**21**	**15**	**0**	–	–

Table 2.13. *Principal machine u*(k)*

$Q^*(0)$	$Q^*(1)$	$Q^*(2)$	$Q^*(3)$	$Q^*(4)$	$Q^*(5)$
25	**25**	**21**	**25**	**0**	**15**
$Q^*(6)$	$Q^*(7)$	$Q^*(8)$	$Q^*(9)$	$Q^*(10)$	$Q^*(11)$
1	**6**	**12**	**18**	**0**	**18**
$Q^*(12)$	$Q^*(13)$	$Q^*(14)$	$Q^*(15)$	$Q^*(16)$	$Q^*(17)$
12	**0**	**20**	**12**	**9**	**16**
$Q^*(18)$	$Q^*(19)$	$Q^*(20)$	$Q^*(21)$	$Q^*(22)$	$Q^*(23)$
14	**4**	**0**	**14**	**11**	**14**
$Q^*(24)$	$Q^*(25)$	$Q^*(26)$	$Q^*(27)$	$Q^*(28)$	$Q^*(29)$
11	**25**	**8**	**4**	**0**	**23**
$Q^*(30)$	$Q^*(31)$	$Q^*(32)$	$Q^*(33)$	–	–
5	**4**	**25**	**0**	–	–

Table 2.14. *Transported quantity: Q*(k)*

Case 3: Delivery Time $\tau = 3$

Cost = 142427mu / N* = 2

$u^*(0)$	$u^*(1)$	$u^*(2)$	$u^*(3)$	$u^*(4)$	$u^*(5)$
9	**22**	**19**	**22**	**22**	**12**
$u^*(6)$	$u^*(7)$	$u^*(8)$	$u^*(9)$	$u^*(10)$	$u^*(11)$
15	**14**	**22**	**0**	**22**	**0**
$u^*(12)$	$u^*(13)$	$u^*(14)$	$u^*(15)$	$u^*(16)$	$u^*(17)$
2	**19**	**3**	**10**	**22**	**0**
$u^*(18)$	$u^*(19)$	$u^*(20)$	$u^*(21)$	$u^*(22)$	$u^*(23)$
11	**22**	**5**	**4**	**15**	**10**
$u^*(24)$	$u^*(25)$	$u^*(26)$	$u^*(27)$	$u^*(28)$	$u^*(29)$
0	**11**	**22**	**5**	**6**	**11**
$u^*(30)$	$u^*(31)$	$u^*(32)$	–	–	–
0	**5**	**9**	–	–	–

Table 2.15. *Principal machine u*(k)*

$Q^*(0)$	$Q^*(1)$	$Q^*(2)$	$Q^*(3)$	$Q^*(4)$	$Q^*(5)$
25	**25**	**25**	**23**	**20**	**13**
$Q^*(6)$	$Q^*(7)$	$Q^*(8)$	$Q^*(9)$	$Q^*(10)$	$Q^*(11)$
0	**13**	**5**	**10**	**16**	**14**
$Q^*(12)$	$Q^*(13)$	$Q^*(14)$	$Q^*(15)$	$Q^*(16)$	$Q^*(17)$
0	**25**	**8**	**3**	**5**	**20**
$Q^*(18)$	$Q^*(19)$	$Q^*(20)$	$Q^*(21)$	$Q^*(22)$	$Q^*(23)$
8	**9**	**9**	**11**	**20**	**14**
$Q^*(24)$	$Q^*(25)$	$Q^*(26)$	$Q^*(27)$	$Q^*(28)$	$Q^*(29)$
0	**6**	**25**	**0**	**11**	**15**
$Q^*(30)$	$Q^*(31)$	$Q^*(32)$	–	–	–
10	**19**	**7**	–	–	–

Table 2.16. *Transported quantity: Q*(k)*

For each value of τ, we determined the economical production plan, the optimal delivery quantity plan and the optimal number of preventive maintenance. We note that for delivery time $\tau = 2$, we obtained the optimal total cost equals to *134987 mu* with two preventive maintenance actions during the horizon $H = 36$, this optimal total cost has become expensive for $\tau = 1$ and equals to *149230 mu* with two preventive maintenance actions and more economy equals *142427 mu* with two preventive maintenance actions for $\tau = 3$. However, in order to find an explanation of this optimal delivery time, we have to discuss the determined optimal production and optimal delivery quantity plans for the deferent delivery time value.

We interpret from the variability of the delivery time that delivery planning changes when the τ changes. For example, if we see in Tables 2.12, 2.14 and 2.16 the quantity to deliver at the period 25 (i.e. $Q^*(25)$), we have $Q^*(25) = 0$ when $\tau = 1$, $Q^*(25) = 25$ when $\tau = 2$ and $Q^*(25) = 6$ when $\tau = 3$. Also, the production planning changes when the τ changes (see Tables 2.11, 2.13 and 2.15). Indeed, this phenomenon is explained by the fact that when the delivery time varies, that means that the arriving of the products from the manufacturing store to the purchase warehouse varies. In other words, more the delivery time increase more the transported products arrive late to the purchase warehouse (i.e. more the purchase warehouse is fulfilled late) and vice-versa. Thus, the delivery time has an impact on the manufacturing store and the purchase warehouse levels (i.e. inventory cost), and then of course it has an impact on the forecasting of the production and delivery quantity planning and also on optimal number of preventive maintenance actions. In conclusion, the optimal planning of production, delivery quantity and maintenance changes according to the delivery time value, and then of course the total cost changes. Thus, through the study of the variability of delivery time, we determined the suitable delivery time value that minimises the total cost and which corresponds to $\tau = 2$ in our numerical example. This study is very useful for one company that should determine the optimal delivery time in order to minimise as possible the total cost, and above all in recent years where the industrial domain has seen a strong competition between companies.

Example 2 (Variability of returned products)

The return of the products to stock is called the withdrawal right. This right gives the customer a specific deadline for returning products. Our contribution is to consider a relation between transportation delay and the rate for returned product. The relation takes into account the minimisation of total cost. This relation correlates the transportation delay and the definition of the withdrawal period. The purpose of this relationship shows that we can find the best optimal combination of cost between the transportation delay and the rate for returned product. This can be explained by the fact that, as the quantity of returned product to the stock increases in order to satisfy the customer and reduce the pressure on the principal and therefore decreases the degradation degree of machine, consequently the optimal number of preventive maintenance actions N. In the opposite case, if the rate of returned product decreases, the quantity of returned product decreases and consequently the principal machine will work more and in this case the degradation degree of machine increases and consequently the number of preventive maintenance actions.

In what follows, we are interested to find the optimal production plan and the optimal delivery quantity plan for each value of returned products rate δ. According to Figures 2.8 and 2.9, we can see that the higher value of returned products rate δ (δ = 0,6) corresponds to the lowest total cost. This can be explained by the fact that, as returned products rate δ increases, normally the inventory level increases and which means that we have enough products in the warehouse S_0 to satisfy the customer. However, when δ increases we need less products which are arriving from the manufacturing store S_1. Therefore, in this case, the number of products produced by machine M and of course the delivery quantities decrease. Furthermore, when the number of products produced by machine M decreases the number of preventive maintenance actions decreases.

According to Figure 2.10, the lowest total cost corresponds to δ = 0,3 (higher value), $\tau = 3$ (higher value) and $N = 2$. However, when δ = 0,1 the total cost is still high for all values of τ and N. This can be clarified by the reality that, as returned products rate increases, the level of the purchase warehouse (from where the customer demand is satisfied) increases, therefore the number of products produced by machine M and the delivery quantities decrease, then consequently the total production/inventory/delivery/maintenance cost decreases.

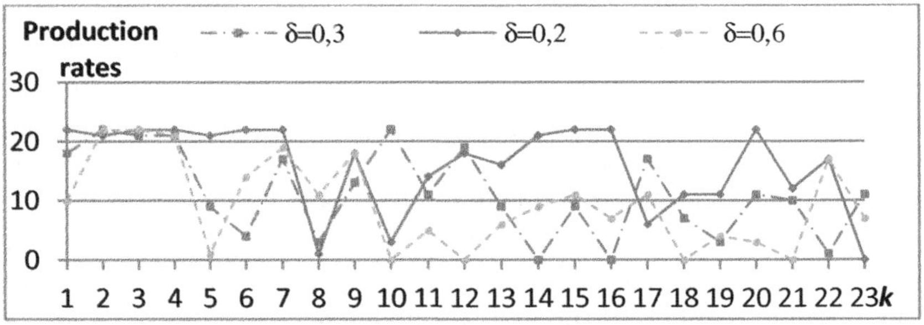

Figure 2.8. *Number of products produced by machine M variation as a function of product returned δ*

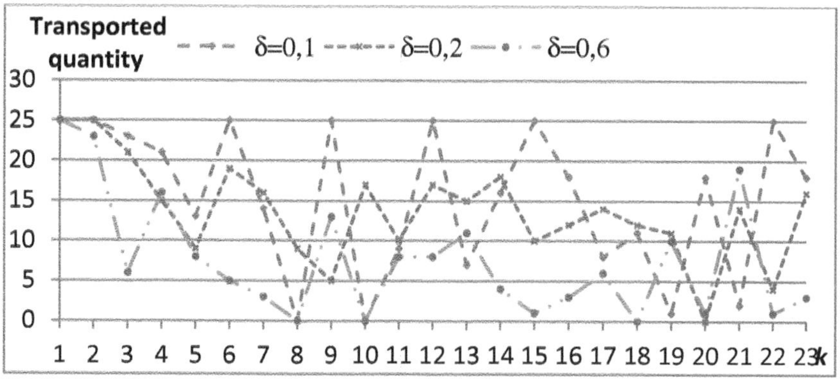

Figure 2.9. *Transported quantity variation as a function of product returned δ*

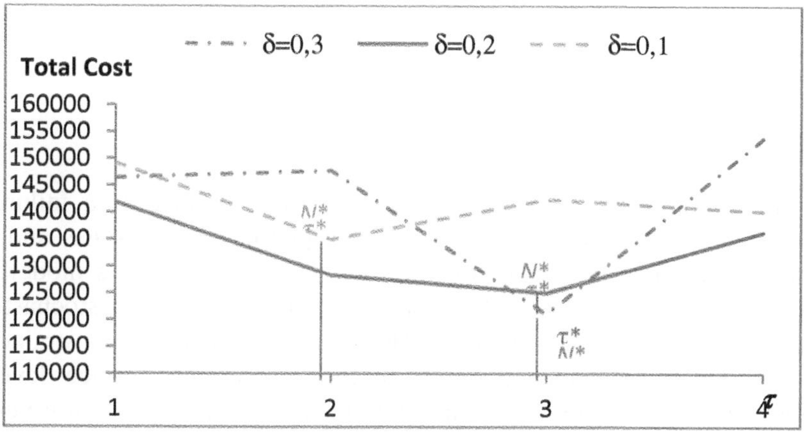

Figure 2.10. *Total cost and delivery time are function of product returned δ*

Example 3 (Variability of service level)

On the other hand, using Figure 2.11, the higher value of the total production/inventory/delivery and maintenance costs corresponds to the higher of the service level. We can note that, as service level α increases, the machine produces more, and then we need to hold longer the products in the manufacturing store and the purchase warehouse in order to meet the demand and to satisfy the higher service level. Consequently, also the transported quantity increases.

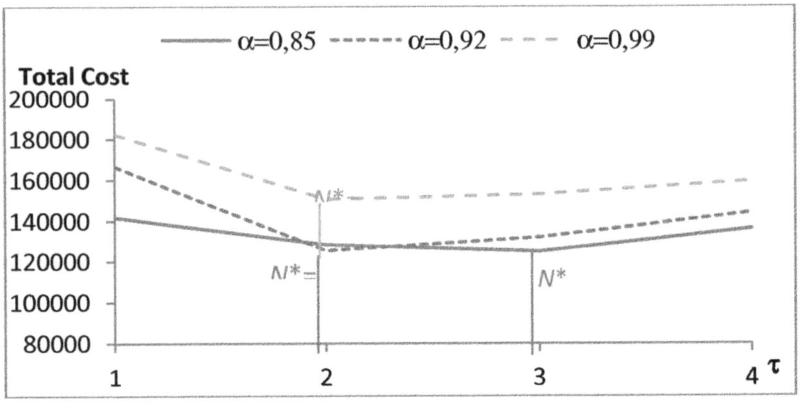

Figure 2.11. *Total cost and delivery time are function of service level α*

2.5. Conclusion

In this chapter, a manufacturing system consisting of a failure-prone machine, a manufacturing store, a purchase warehouse, and a transport vehicle has been considered. The delivery time is modeled taking into account vehicle capacity, machine failures, stochastic demand and product returns. This work studied a constrained production–maintenance–delivery problem including service level, stochastic demand, delivery time, failure rate and product return rate. A preventive maintenance plan is provided in order to decrease the failure rate. A stochastic delivery–maintenance–production problem that is based on a quadratic model is formulated and then solved. Taking into account the influence of delivery time, we have sequentially optimized the problem to obtain optimal delivery and production plans with the corresponding preventive maintenance periods. Numerical examples have been presented which illustrate the proposed approach and the robustness of our model. The variability in delivery time is elaborated on in order to study the influence of delivery time on optimal production and the delivery plans. Variability in the rates of product return and service level are studied in order to examine their impact on optimal production and delivery plans. Furthermore, we study an equivalent model, which considers that the rate at which a product is returned can be subtracted directly from demand. The results of the variability study show that when the rate of product return increases, the number of products produced by machine M and the quantities of products delivered decrease in response. For service level, when the service level increases, the number of products produced by machine M and the quantities delivered increase.

For future research, we will consider a more complex system with one manufacturing store, multiple warehouses and multiple transport vehicles. We will also consider more than one customer, each with stochastic demands.

3

A Decision Optimization Model for Leased Manufacturing Equipment with Warranty for a Production–Maintenance Forecasting Problem

Due to the cost of equipment used in industrial production, many manufacturers lease equipment with a warranty period over a finite leasing horizon, rather than purchasing it. The lease contains the possibility of obtaining an extended warranty at an additional cost. In this chapter, based on a production–maintenance forecasting optimization problem, we develop a mathematical model to study the lease contract with a basic and extended warranty based on a win–win relationship for the lessee and the lessor. The influence of the rates of production on equipment degradation and consequently on the total cost for each side over the finite leasing horizon is stated in order to determine a theoretical condition in which a compromise-pricing zone exists under different maintenance policies.

3.1. Introduction

Due to rapid advances in technology in recent decades, the creation of new and better equipment has led to a high rate of technological obsolescence in the market. The cost of owning new equipment is very high so more industries have started leasing equipment rather than buying it. When agreeing a leasing contract, the number and frequency of maintenance actions are the most important element in negotiations between the manufacturer and the consumer.

In the past, when businesses owned equipment, the different preventive and corrective maintenance actions were approved internally. This began to change with the development of increasingly complex equipment to carry out specialist services. The maintenance of such equipment rapidly became uneconomical to carry out in-house.

Most manufacturers do not consider maintenance a basic activity, which is why they focus only on the principal activities, which are considered the basic tenets of the business. In this context, the notion of a warranty is attached to the concept of leasing equipment, since the contract contains the details and guarantees the maintenance service.

In light of these developments and the reasoning behind them, we can consider the maintenance warranty to be an attractive selling point when leasing equipment as it increases manufacturer (lessee) confidence, thus increasing the interest of potential lessors. From the customer's (lessee's) point of view, having a warranty means the cost of repairs or replacement of defective equipment during the warranty period is reduced.

The maintenance of leased equipment offered by the lessor is generally specified in a lease contract agreed by the lessor to the lessee [MUR 99]. Some research has analyzed the problem of maintaining leased equipment, and numerous preventive maintenance policies, which have been proposed and studied in various situations, such as perfect or imperfect maintenance. Yeh *et al.* [YEH 07] determine the optimal number of lease periods and define the maintenance strategy for leased equipment based on a model of minimal repair to restore the equipment to an operating condition when it fails and imperfect preventive maintenance to avoid failure until the age of the equipment reaches a certain threshold value. In the same context, Jaturonnatee *et al.* [JAT 06] proposed a method in which the rate at which equipment fails is reduced after each preventive maintenance action; they solve the optimal maintenance policy of leased equipment under periodical preventive maintenance actions.

Due to the complexity of maintenance actions required for various machines, lessees prefer to pay for an extended warranty period at extra cost in order to avoid the problems of increased production–maintenance costs and production system perturbation. In this context, Berke and Zaino [BER 91] treated two types of contract that were intended to assure the lessee that the product would perform its planned functions under specific conditions and for specified periods of time. The first type of contract defined a combination policy that proposed an initial free replacement warranty and from a certain period the replacement item's cost is calculated on a sliding scale. The second type of contract was the fleet warranty, which guarantees the purchaser of a large quantity of items an average field performance.

Continuing to focus on the aspect of warranty and maintenance, Kim *et al.* [KIM 04] defined the relation between the warranty and preventive maintenance by showing the impact of preventative maintenance over the warranty period on the cost of warranty service. Won *et al.* [WON 08] proposed two new warranty servicing strategies, concerning imperfect and minimal repairs. In the first strategy, they used functional optimization to determine the optimal improvement in reliability when an imperfect repair was carried out during the warranty period and dependent upon the age of the item. In the second strategy, they include only an optimization parameter to determine the optimal improvement in reliability that does not depend on the age of the item being leased.

There are other types of warranties that can be applied to non-repairable products, such as the renewing free-replacement warranty (RFRW) where, in the case of product failure during the warranty period, the product is replaced with a new one with a full warranty. Chien *et al.* [CHI 08] analytically investigated the impact of the RFRW on the optimal age replacement policy for a repairable product with a general failure model. They presented a general model containing two types of failure when the product fails:

– type 1: minor failure, in which the failure is addressed by a minimal repair;

– type 2: catastrophic failure, which can only be remedied by a replacement.

In another paper, Chien *et al.* [CHI 10] presented a new warranty strategy based on an age-replacement policy for products. This strategy combined a fully RFRW with a pro-rata warranty policy (RFRW/RPRW policy). They developed a cost model from the user/buyer perspective and discussed special cases in order to determine the corresponding local optimal replacement age by minimizing the expected cost rate of a long run.

Most research concerning warranty problems considers a fixed warranty period, while the dynamic or the extended warranty period – especially in the lease contract – helps the lessor to remain in contact with clients after the end of warranty period. Extended warranty helps the customer (lessee) to continue to receive the same maintenance service for equipment that is well known to the lessor. In this context, Bouguerra *et al.* [BOU 12] developed a mathematical model to study the opportunity provided by the extended warranty for the lessor and lessee and proposed a long guarantee plan consisting of preventive maintenance for systems subjected to random troubleshooting. This strategy considers diverse options for maintenance policies during the following periods:

– basic guarantee period;

– extended guarantee period;

– post guarantee period.

Shaomin and Phil [SHA 11] showed the influence of both the length of warranty period and replacement time on the lifecycle cost of the equipment. They formulated the expected lifecycle cost considering the opportunity-based age replacement policy with minimal repair for an extended warranty and maintenance. They also proved the conditions for the existence of optimal solutions for both the length of the extended warranty period and the design life in special cases.

Recently, another type of problem that deals with leasing/warranty problem has been treated in [HAJ 13]. Hajej *et al.* handled the optimization problem of production and maintenance policies for equipment under a lease contract with warranty periods. A mathematical model of the total production and maintenance costs is developed and optimal production planning as well as the corresponding optimal maintenance strategy is derived by choosing the optimal warranty periods for the lessee in order to minimize the total cost.

Motivated by Hajej *et al.* [HAJ 13], we can consider that work as a continuation of our work determining the most optimal basic warranty periods for the lessee. This study shows novelty and originality relative to this type of problem, which uses a mathematical model to study the opportunity provided by extending the warranty for the lessee. Based on a production and maintenance forecasting problem for a leased machine, we will determine the total cost of for each side in order to determine, for any given situation, the area of possible compromise yielding a win–win relationship with respect to the extended warranty cost. The area of compromise is characterized by the maximum extra cost the lessee should pay for the extended warranty, and the minimum price at which the lessor should sell it. We will show the influence of production rates as well as maintenance actions on the manufacturing machine on the servicing cost over the warranty and extended warranty periods.

This study proposes the idea of production and maintenance coupling in a lease with a warranty. It has novelty and originality relative to this type of problem, as it considers and proposes a new maintenance strategy for a leasing contract with extended warranty based on a win–win relationship between the lessee and the lessor. This is characterized by the influence of the variation in production rates on the degree of machine degradation characterized by analytical study that shows evolution of the machine failure rate according to its use, at the same time respecting the continuity of equipment reliability between one period and another. Secondly, in our opinion, no analytical or numerical model has been stated in the literature that leads to a decision framework where the lessee and/or lessor identify extended warranty pricing zones that are acceptable for both sides.

The remainder of this chapter is organized as follows:

– section 3.2 states the problem;

– section 3.3 presents and develops the mathematical model concerning the production forecasting problem and the different maintenance policies considering the influence of production rates on the degradation of the leasing machine;

– section 3.4 presents a numerical example illustrating our approach, followed by a variability study showing the impact of variation in preventive maintenance costs on our model;

– the conclusion is given in section 3.5.

3.2. Description of the problem

3.2.1. *Notation*

We have used the following notations in this chapter:

Δt	length of a production period
L	number of leasing periods
X	warranty period
Xe	warranty period including both basic period X and extension
U_k	production rate of machine M during period k ($k = 0, 1, \ldots, L$)
$\hat{d}(k)$	average demand during period k ($k = 0, 1, \ldots, L$)
$V_{d(k)}$	variation in demand during period k ($k = 0, 1, \ldots, L$)

S_k	inventory level of S at the end of period k ($k = 0, 1, ..., L$)
$\hat{S}_k$	average inventory level of S during period k ($k = 0, 1, ..., L$)
C_{pr}	unit production cost of the leasing machine
Cs	holding cost of a unit of product during one period
mu	monetary unit
U^{max}	maximum production rate of the leasing machine
U^{min}	minimum production rate of the leasing machine
α	probability index related to customer satisfaction and expressing the service level
S_0	initial inventory
Cpm	preventive maintenance cost
Ccm	corrective maintenance cost

3.2.2. *Statement of the problem*

In this chapter, we consider a production and maintenance forecasting problem for a leased machine with warranty periods. We aim to define a new aspect of the leasing contract. Generally, several pieces of equipment are leased with a warranty period but there are leasing contracts that for an additional cost give the lessee the possibility of purchasing an additional period of warranty, which will start at the end of the basic warranty period. Hence, the lessee has to decide whether or not to buy the extended warranty and whether or not the price is acceptable. It is a difficult decision for both sides. The lessee does not know if the extra cost (the price of the extended warranty) involved in leasing the equipment will exceed the potential cost of repairs that would be borne by him/her if he/she does not take the extended warranty. On the other hand, to ensure that the lessor does not lose out, the price of the extended warranty should be higher than the cost of the servicing claims (maintenance actions) borne by him/her during the additional warranty period.

We will address all of these issues by proposing a forecasting model in which a manufacturing machine is leased for a multi-horizon, $L .\Delta t$

(we assume that the production horizon is split into equal periods with a length equal to Δt), with multiple periods of warranty, $X \cdot \Delta t$. We suppose that the machine being leased is designed to produce only one type of product in a manufacturing system consisting of a manufacturing store, where the customer receives products over the finite leasing horizon *L*. Moreover, for the forecasting problem, we assume that satisfaction of demand is under a given inventory service level, α, and the fluctuation in demand has a normal distribution with mean and variance given by $\hat{d}$ and σ_d, respectively.

The leased machine considered is subject to random failures. Its failure rate, $\lambda(t)$, increases with both time and production rate. The variation in production rate therefore has an impact on equipment degradation and hence on the average number of failures. This impact is considered in the model.

The leasing contract includes the machine under the warranty period *X* with the possibility of extension until instant *Xe* for an additional cost C_X paid by the lessee when initially leasing the machine; namely, that all maintenance actions during the basic and extended warranty periods are supported by the lessor at no cost to the lessee. For the rest of the leasing periods, the equipment is not under warranty and the maintenance actions are the responsibility of the lessee.

This model uses the preventive maintenance policy with minimal repair at failure with negligible duration, keeping the system failure rate at nearly the same level across the horizon considered. The role of maintenance is to increase the availability of the machine while reducing the maintenance costs in order to ensure the production plan is met across the leasing horizon *L*.

According to the forecasting problem, as well as obtaining the optimal production plan for the machine being leased, our objective is to develop a mathematical model to study the opportunity provided by the extended warranty from the perspectives of the lessee and the lessor. We will express the total expected cost incurred by each side during the product's lifecycle in order to determine, for any given situation, the maximum extra cost the lessee should pay for the extended warranty, and the minimum price at which the lessor should sell it. Taking into account the influence of preventive maintenance actions performed on the leasing machine during the basic and extended warranty periods, we consider different maintenance strategies during the lifecycle of the leased machine.

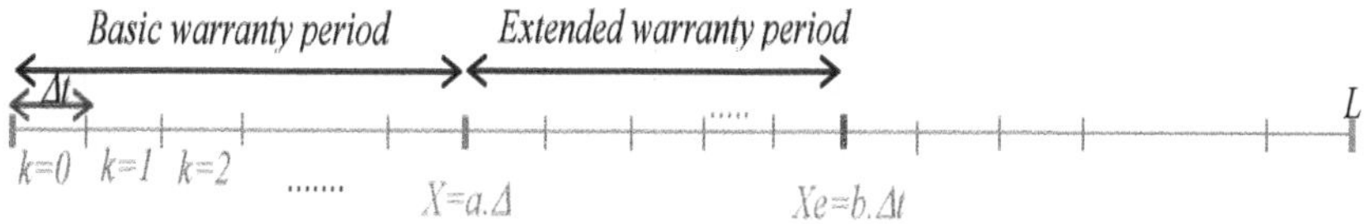

Figure 3.1. *Lifecycle of a leased machine with warranty and extended warranty periods*

3.3. Mathematical model

3.3.1. *Forecast production plan*

3.3.1.1. *Stochastic production model*

Based on the approach proposed by Hajej *et al.* in [HAJ 11] and [HAJ 13], the production planning problem is formulated as a quadratic model, whose decision variables include production rates and inventory levels. The purpose of this section is to develop a mathematical model that will allow us to determine the optimal production plan, U^* ($U^* = u(k)^*$, with and $k = 1,.., L-1$) during the leasing horizon L.

Formally, the stochastic production model is defined as follows:

$$Minimize\ Z\ \sum_{k=0}^{L} f_k(U_k, S_k) = C_s \cdot E\{S_L^2\} + \sum_{k=0}^{L} C_s \cdot E\{S_k^2\} + C_{pr} \cdot U_k^2 \quad [3.1]$$

subject to:

– inventory balance equation constraints

$$S_{k+1} = S_k + U_k - d_k \ k \in \{0,1, \ldots, L-1\} \quad [3.2]$$

– the service level requirement for each period

$$Prob[S_{k+1} \geq 0] \geq \alpha \ k \in \{0,1, \dots, L-1\} \quad [3.3]$$

– capacity constraints

$$0 \leq U_k \leq U^{max} \ k \in \{0,1, \dots, L-1\} \quad [3.4]$$

3.3.1.2. *Deterministic production model*

An approach that transforms the stochastic problem into a deterministic equivalent is necessary. This deterministic problem maintains the main properties of the original problem.

The quadratic total expected cost of production and inventory over the leasing periods can then be expressed as follows:

$$Z(u) = C_s \times (\hat{S}_L^{\,2}) + \sum_{k=0}^{L-1} C_s \cdot \hat{S}_k^{\,2} + C_{pr} \times u_k^{\,2} + C_s \times \sigma_d^{\,2} \times \frac{L(L+1)}{2} \quad [3.5]$$

with:

– mean variables: $E\{S_k\} = \hat{S}_k$;

– variance variables: $E\{u_k\} = u_k$ (variable u_k is deterministic).

The inventory balance equation [3.2] can be reformulated as:

$$\hat{S}_{k+1} = \hat{S}_k + u_k - \hat{d}_k \quad [3.6]$$

PROOF.– See Appendix 2.

3.3.1.3. *Service level constraint*

Taking another step to transform the stochastic problem into an equivalent deterministic one, we consider a service level constraint in a deterministic form by determining the minimum cumulative quantity produced depending on the service level requirements.

For $k \in \{0,1, \dots, h_i - 1\}$, we have:

$$\Pr ob\left(S_{k+1} \geq 0\right) \geq \alpha \Rightarrow \left(U_k \geq V_{d,k} \cdot \varphi_{d,k}^{-1}(\alpha) + \hat{d}_k - \hat{S}_k\right) \quad [3.7]$$

where:

– $\varphi_{d,k}$ is the cumulative Gaussian distribution function with mean $\hat{d}_k$ and finite variance $Var(d_k) = V_{d,k} \geq 0$;

– $V_{d,k}$ is the variation in demand d during period k.

PROOF.– See Appendix 2.

3.3.2. *Maintenance policy*

Based on Hajej *et al.* [HAJ 11], the maintenance strategy considers the manufacturing system's degradation according to the production rate during the leasing horizon L. The correlation between the degradation of the machine and production rates is manifested by an increased failure rate with increased time and production rate.

We assume that during the machine's lifecycle, perfect preventive maintenance or replacement are performed periodically at times $i.T$, $i = 0,1,\ldots, N_j$ (with N_j being the number of preventive maintenance actions over each leasing period: basic warranty, extended warranty, post warranty; and where T is the preventive maintenance action interval) following which the unit is as good as new.

The evolution of the machine's failure rate according to its use (which in our case is the production rate for each period) respects the continuity of equipment reliability from one period to another. This is presented by an analytical equation.

The failure rate in interval k is expressed as follows:

$$\lambda_k(t) = \lambda_{k-1}(\Delta t) + \frac{U_k}{U_{max}} \cdot \lambda_n(t) \ \forall\, t \in [0, \Delta t] \qquad [3.8]$$

with:

$$\lambda_{k=0} = \lambda_0 \text{ and } \Delta\lambda_k(t) = \frac{U_k}{U_{max}} \cdot \lambda_n(t) \qquad [3.9]$$

where $\lambda_n(t)$ is the nominal failure rate corresponding to the maximal production rate.

According to the maintenance strategy, we can define the different numbers of preventive maintenance actions over each interval during the leasing periods given as follows:

– N_1: number of preventive maintenance actions during the basic warranty periods $[0, X)$ with a value equal to $In\left(\frac{X}{T}\right)$;

– N_2: number of preventive maintenance actions during the leasing periods [0, L) with a value equal to $In\left(\frac{L}{T}\right)$;

– N_3: number of preventive maintenance actions between the end of basic warranty and the end of the leasing periods $[X, L)$ with a value equal to $In\left(\frac{L-X}{T}\right)$;

– N_4: number of preventive maintenance actions during the basic and extended warranty periods $[0, X_e)$ with a value equal to $In\left(\frac{X_e}{T}\right)$;

– N_5: number of preventive maintenance actions during the extended warranty $[X, Xe)$ with a value equal to $In\left(\frac{Xe-X}{T}\right)$;

– N_6: number of preventive maintenance actions between the end of the extended warranty periods and the end of leasing periods $[X_e$, L) with a value equal to $In\left(\frac{L-Xe}{T}\right)$, where In is the integer part of a real number.

The analytic expression of the total maintenance cost incurred by each side during the leasing period where $\varphi_M(U, N_i)$ corresponds to the expected number of failures that occur during the different intervals defined above, considering the production rate in each production period Δt, is:

$$\xi(U, N_i) = C_{pm} \times (N_i - 1) + C_{cm} \times \varphi_M(U, N_i) \qquad [3.10]$$

Let *In* denote the integer part of (.). The average number of machine failures rate defined above is therefore:

$$\varphi_M(U,N_i)=\sum_{j=0}^{N_i-1}\left[\sum_{i=In(j\times\frac{T}{\Delta t})+1}^{In\left((j+1)\times\frac{T}{\Delta t}\right)}\int_0^{\Delta t}\lambda_i(t)+\int_0^{(j+1)\times T-In\left((j+1)\times\frac{T}{\Delta t}\right)\times\Delta t}\lambda_{In\left((j+1)\times\frac{T}{\Delta t}\right)+1}(t)dt\right.$$
$$\left.+\int_{(j+1)\times T}^{\left(In\left((j+1)\times\frac{T}{\Delta t}\right)+1\right)\times\Delta t}\frac{\left(\left(In\left((j+1)\times\frac{T}{\Delta t}\right)+1\right)\right)}{U_{\max}}\times\lambda_n(t)dt\right]\quad[3.11]$$

Using the total cost, we can determine the area of possible compromise yielding a win–win relationship characterized by the maximum additional cost the lessee should pay for the extended warranty, and the minimum price at which the lessor should sell it. There are different maintenance strategies that can be adopted across the leasing horizon, taking into account the impact of preventive maintenance on the warranty servicing cost. The following maintenance policies will be considered from lessor and lessee perspectives:

– *Policy I*: periodic preventive maintenance actions during the post basic warranty period. For this policy we consider the following possibility:

- *Policy I-1*: preventive maintenance actions are performed during the extended warranty [X, *Xe*) at times $i.T$, i = 0,1,…,N_5 supported by the lessor, and such actions are performed from the end of the extended warranty [*Xe*, *L*) at times $i.T$, i = 0,1,…,N_6 supported by the lessee;

– *Policy II*: periodic preventive maintenance actions during the warranty period. For this case we consider two different possibilities:

- *Policy II-1*: preventive maintenance actions are performed only during the basic warranty period [0, X) at times $i.T$, i = 0,1,…,N_1 supported by the lessor,

- *Policy II-2*: preventive maintenance actions are performed during both the basic and the extended warranty periods $[0, X_e)$ at times $i.T$, $i = 0,1,\ldots,N_4$ supported by the lessor.

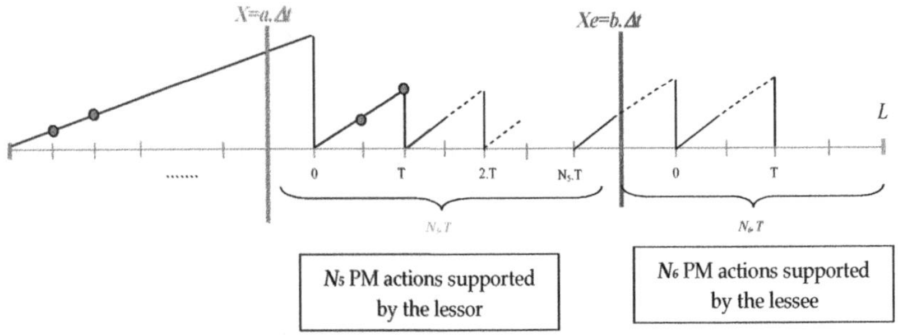

Figure 3.2. *Evolution of the rate of failure for policy I.1*

3.3.3. *Maximum additional cost for an extended warranty*

Here we determine the maximum additional cost that the lessee should pay for the extended warranty during the leasing periods. Comparing the total maintenance costs the lessee is liable for if he/she takes or does not take the extended warranty period requires us to determine the cost of the extended warranty period paid by the lessee. For each maintenance policy, the best situation when buying an extended warranty period would be when it costs the lessee the least amount of money, i.e. when the total maintenance cost incurred during the leasing horizon is lower than it would cost to maintain the machine(s) if she/she does not take it.

We assume that ξ_{cPn} and ξ_{cPy} are the total maintenance costs covered by the lessee's maintenance policy (P) in two situations: when he does not take the extended warranty period (n), and when he does take it (y).

We recall that:

$$N_1 = \frac{X}{T}$$

$$N_3 = \frac{L-X}{T}$$

$$N_5 = \frac{X_e - X}{T}$$

$$N_6 = \frac{L - X_e}{T}$$

$$X = a \cdot \Delta t$$

and

$$X_e = a_1 \cdot \Delta t$$

where Δt is the length of the production period.

– *Policy I*:

- *Policy I-1*:

$$\xi_{cI-1y} + \xi_{cX} \leq \xi_{cI-1n} \Rightarrow \xi_{cX} \leq \xi_{cI-1n} - \xi_{cI-1y} \Rightarrow \xi_{cX} \leq A_{cI}$$

where:

$$\xi_{cI-1n} = C_{cm} \times \left[\sum_{i=1}^{In\left(\frac{X}{\Delta t}\right)} \int_0^{\Delta t} \lambda_i(t)dt + \sum_{j=0}^{N_3-1} \sum_{i=In\left(j \cdot \frac{T}{\Delta t}\right)+\frac{X}{\Delta t}+1}^{In\left((j+1)\cdot\frac{T}{\Delta t}\right)+\frac{X}{\Delta t}} \int_0^{\Delta t} \lambda_i(t)dt + \sum_{i=In\left(N_3 \cdot \frac{T}{\Delta t}\right)+\frac{X}{\Delta t}+1}^{\frac{L}{\Delta t}} \int_0^{\Delta t} \lambda_i(t)dt \right] + N_3 \times C_{pm}$$

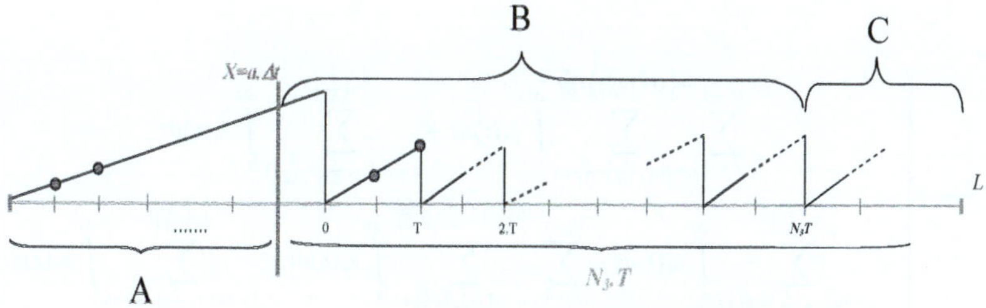

Figure 3.3. *Average number of failures when the lessee does not extend the warranty period*

$$\xi_{cI-1y}\ C_{cm}\left[\sum_{i=1}^{In\left(\frac{X}{\Delta t}\right)}\int_{0}^{\Delta t}\lambda_i(t)dt + \sum_{i=In\left(N_5\cdot\frac{T}{\Delta t}\right)+\frac{X}{\Delta t}+1}^{In\left(\frac{X_e}{\Delta t}\right)}\int_{0}^{\Delta t}\lambda_i(t)dt + \sum_{j=0}^{N_6-1}\sum_{i=In\left(j\cdot\frac{T}{\Delta t}\right)+\frac{X_e}{\Delta t}+1}^{In\left((j+1)\cdot\frac{T}{\Delta t}\right)+\frac{X_e}{\Delta t}}\int_{0}^{\Delta t}\lambda_i(t)dt + \sum_{i=In\left(N_6\cdot\frac{T}{\Delta t}\right)+\frac{X_e}{\Delta t}+1}^{\frac{L}{\Delta t}}\int_{0}^{\Delta t}\lambda_i(t)dt\right] + N_6\times C_{pm}$$

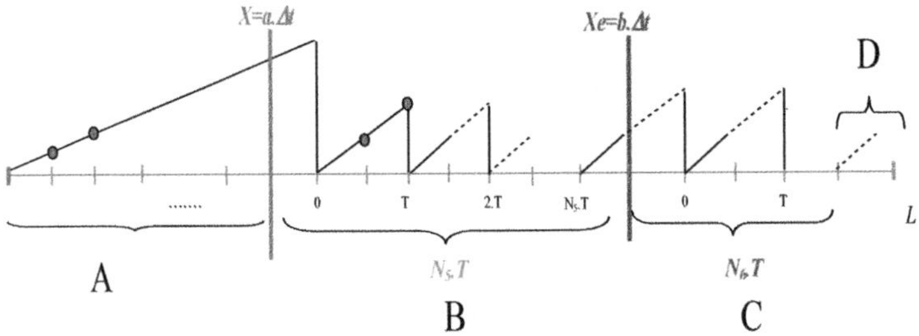

Figure 3.4. *Average number of failures if the lessee does extend the warranty period*

$$A_{cI} = C_{cm}\times\left[\sum_{j=0}^{N_3-1}\sum_{i=In\left(j\cdot\frac{T}{\Delta t}\right)+\frac{X}{\Delta t}+1}^{In\left((j+1)\cdot\frac{T}{\Delta t}\right)+\frac{X}{\Delta t}}\int_{0}^{\Delta t}\lambda_i(t)dt + \sum_{i=In\left(N_3\cdot\frac{T}{\Delta t}\right)+\frac{X}{\Delta t}+1}^{In\left(\frac{L}{\Delta t}\right)}\int_{0}^{\Delta t}\lambda_i(t)dt - \sum_{i=In\left(N_5\cdot\frac{T}{\Delta t}\right)+\frac{X}{\Delta t}+1}^{In\left(\frac{X_e}{\Delta t}\right)}\int_{0}^{\Delta t}\lambda_i(t)dt - \sum_{j=0}^{N_6-1}\sum_{i=In\left(j\cdot\frac{T}{\Delta t}\right)+\frac{X_e}{\Delta t}+1}^{In\left((j+1)\cdot\frac{T}{\Delta t}\right)+\frac{X_e}{\Delta t}}\int_{0}^{\Delta t}\lambda_i(t)dt - \sum_{i=In\left(N_6\cdot\frac{T}{\Delta t}\right)+\frac{X_e}{\Delta t}+1}^{In\left(\frac{L}{\Delta t}\right)}\int_{0}^{\Delta t}\lambda_i(t)dt\right] + C_{pm}\times(N_3-N_6) \qquad [3.12]$$

– *Policy II*:

- *Policy II-1*:

$$\xi_{cII-1y} + \xi_{cw} \leq \xi_{cII-1n} \rightarrow \xi_{cX} \leq \xi_{cII-1n} - \xi_{cII-1y} \rightarrow \xi_{cX} \leq B_{c1}$$

$$\begin{aligned}\xi_{cII-1n} = C_{cm}\Bigg[& \sum_{i=In\left(N_1 \cdot \frac{T}{\Delta t}\right)+1}^{In\left(\frac{X}{\Delta t}\right)} \int_0^{\Delta t} \lambda_i(t)dt \\ & + \sum_{j=0}^{N_3-1} \sum_{i=In\left(j\cdot\frac{T}{\Delta t}\right)+\frac{X}{\Delta t}+1}^{In\left((j+1)\cdot\frac{T}{\Delta t}\right)+\frac{X}{\Delta t}} \int_0^{\Delta t} \lambda_i(t)dt + \sum_{i=In\left(N_3\cdot\frac{T}{\Delta t}\right)+\frac{X}{\Delta t}+1}^{In\left(\frac{L}{\Delta t}\right)} \int_0^{\Delta t} \lambda_i(t)dt\Bigg] \\ & + N_3 \times C_{pm}\end{aligned}$$

$$\begin{aligned}\xi_{cII-1y} = C_{cm}\Bigg[& \sum_{i=In\left(N_1 \cdot \frac{T}{\Delta t}\right)+1}^{In\left(\frac{X}{\Delta t}\right)} \int_0^{\Delta t} \lambda_i(t)dt + \sum_{i=\frac{W}{\Delta t}}^{\frac{X_e}{\Delta t}} \int_0^{\Delta t} \lambda_i(t)dt \\ & + \sum_{j=0}^{N_6-1} \sum_{i=In\left(j\cdot\frac{T}{\Delta t}\right)+\frac{X_e}{\Delta t}+1}^{In\left((j+1)\cdot\frac{T}{\Delta t}\right)+\frac{X_e}{\Delta t}} \int_0^{\Delta t} \lambda_i(t)dt + \sum_{i=In\left(N_6\cdot\frac{T}{\Delta t}\right)+\frac{X_e}{\Delta t}+1}^{In\left(\frac{L}{\Delta t}\right)} \int_0^{\Delta t} \lambda_i(t)dt\Bigg] \\ & + N_6 \times C_{pm}\end{aligned}$$

$$\begin{aligned}B_{c1} = C_{cm}\Bigg[& \sum_{j=0}^{N_3-1} \sum_{i=In\left(j\cdot\frac{T}{\Delta t}\right)+\frac{X}{\Delta t}+1}^{In\left((j+1)\cdot\frac{T}{\Delta t}\right)+\frac{X}{\Delta t}} \int_0^{\Delta t} \lambda_i(t)dt \\ & + \sum_{i=In\left(N_3\cdot\frac{T}{\Delta t}\right)+\frac{X}{\Delta t}+1}^{In\left(\frac{L}{\Delta t}\right)} \int_0^{\Delta t} \lambda_i(t)dt - \sum_{i=In\left(\frac{X}{\Delta t}\right)}^{In\left(\frac{X_e}{\Delta t}\right)} \int_0^{\Delta t} \lambda_i(t)dt \\ & - \sum_{j=0}^{N_6-1} \sum_{i=In\left(j\cdot\frac{T}{\Delta t}\right)+\frac{X_e}{\Delta t}+1}^{In\left((j+1)\cdot\frac{T}{\Delta t}\right)+\frac{X_e}{\Delta t}} \int_0^{\Delta t} \lambda_i(t)dt - \sum_{i=In\left(N_6\cdot\frac{T}{\Delta t}\right)+\frac{X_e}{\Delta t}+1}^{In\left(\frac{L}{\Delta t}\right)} \int_0^{\Delta t} \lambda_i(t)dt\Bigg] \\ & + C_{pm} \times (N_3 - N_6)\end{aligned}$$

[3.13]

- *Policy II-2*:

$$\xi_{cII-2y} + \xi_{cX} \leq \xi_{cII-2n} \rightarrow \xi_{cX} \leq \xi_{cII-2n} - \xi_{cII-2y} \rightarrow \xi_{cX} \leq B_{c2}$$

$$\xi_{cII-2n} = C_{cm}\left[\sum_{i=In\left(N_1\cdot\frac{T}{\Delta t}\right)+1}^{In\left(\frac{X}{\Delta t}\right)} \int_0^{\Delta t} \lambda_i(t)dt + \sum_{j=0}^{N_3-1} \sum_{i=In\left(j\cdot\frac{T}{\Delta t}\right)+\frac{X}{\Delta t}+1}^{In\left((j+1)\cdot\frac{T}{\Delta t}\right)+\frac{X}{\Delta t}} \int_0^{\Delta t} \lambda_i(t)dt + \sum_{i=In\left(N_3\cdot\frac{T}{\Delta t}\right)+\frac{X}{\Delta t}+1}^{In\left(\frac{L}{\Delta t}\right)} \int_0^{\Delta t} \lambda_i(t)dt\right] + N_3 \times C_{pm}$$

$$\xi_{cII-2y} = C_{cm}\left[\sum_{i=In\left(N_5\cdot\frac{T}{\Delta t}\right)+\frac{X}{\Delta t}+1}^{In\left(\frac{X_e}{\Delta t}\right)} \int_0^{\Delta t} \lambda_i(t)dt + \sum_{j=0}^{N_6-1} \sum_{i=In\left(j\cdot\frac{T}{\Delta t}\right)+\frac{X_e}{\Delta t}+1}^{In\left((j+1)\cdot\frac{T}{\Delta t}\right)+\frac{X_e}{\Delta t}} \int_0^{\Delta t} \lambda_i(t)dt + \sum_{i=In\left(N_6\cdot\frac{T}{\Delta t}\right)+\frac{X_e}{\Delta t}+1}^{In\left(\frac{L}{\Delta t}\right)} \int_0^{\Delta t} \lambda_i(t)dt\right] + N_6 \times C_{pm}$$

$$B_{c2} = C_{cm}\left[\sum_{i=In\left(N_1\cdot\frac{T}{\Delta t}\right)+1}^{In\left(\frac{X}{\Delta t}\right)} \int_0^{\Delta t} \lambda_i(t)dt + \sum_{j=0}^{N_3-1} \sum_{i=In\left(j\cdot\frac{T}{\Delta t}\right)+\frac{X}{\Delta t}+1}^{In\left((j+1)\cdot\frac{T}{\Delta t}\right)+\frac{X}{\Delta t}} \int_0^{\Delta t} \lambda_i(t)dt + \sum_{i=In\left(N_3\cdot\frac{T}{\Delta t}\right)+\frac{X}{\Delta t}+1}^{In\left(\frac{L}{\Delta t}\right)} \int_0^{\Delta t} \lambda_i(t)dt - \sum_{i=In\left(N_5\cdot\frac{T}{\Delta t}\right)+\frac{X}{\Delta t}+1}^{In\left(\frac{X_e}{\Delta t}\right)} \int_0^{\Delta t} \lambda_i(t)dt - \sum_{j=0}^{N_6-1} \sum_{i=In\left(j\cdot\frac{T}{\Delta t}\right)+\frac{X_e}{\Delta t}+1}^{In\left((j+1)\cdot\frac{T}{\Delta t}\right)+\frac{X_e}{\Delta t}} \int_0^{\Delta t} \lambda_i(t)dt - \sum_{i=In\left(N_6\cdot\frac{T}{\Delta t}\right)+\frac{X_e}{\Delta t}+1}^{In\left(\frac{L}{\Delta t}\right)} \int_0^{\Delta t} \lambda_i(t)dt\right] + C_{pm} \times (N_3 - N_6)$$

[3.14]

3.3.4. *Minimum price at which to sell the extended warranty*

Here we determine the minimum price at which the lessor can sell the extended warranty during leasing periods. We establish, for each maintenance policy, the best situation for the lessor, i.e. the lessor wins as it costs him/her less to maintain the machine(s) than the amount paid by the lessee for the extended warranty.

We assume that ξ_{MPn} and ξ_{MPy} are the total maintenance costs the lessor charges for the maintenance policy (P) in two scenarios: without the extended warranty period (n) and with the extended warranty period (y).

– *Policy I*:

- *Policy I-1*:

$$\xi_{MI-1y} - \xi_{MX} \leq \xi_{MI-1n} \rightarrow \xi_{MX} \geq \xi_{MI-1y} - \xi_{MI-1n} \rightarrow \xi_{MX} \geq A_{M1}$$

$$\xi_{MI-1n} = C_{cm} \times \left[\sum_{i=1}^{In(\frac{X}{\Delta t})} \int_0^{\Delta t} \lambda_i(t)dt \right]$$

$$\xi_{MI-1y} = C_{cm} \times \left[\sum_{i=1}^{In(\frac{X}{\Delta t})} \int_0^{\Delta t} \lambda_i(t)dt + \sum_{j=0}^{N_5-1} \sum_{i=In\left(j\cdot\frac{T}{\Delta t}\right)+\frac{X}{\Delta t}+1}^{In\left((j+1)\cdot\frac{T}{\Delta t}\right)+\frac{X}{\Delta t}} \int_0^{\Delta t} \lambda_i(t)dt \right] + C_{pm} \times N_5$$

$$A_{M1} = C_{cm} \times \left[\sum_{j=0}^{N_5-1} \sum_{i=In\left(j\cdot\frac{T}{\Delta t}\right)+\frac{X}{\Delta t}+1}^{In\left((j+1)\cdot\frac{T}{\Delta t}\right)+\frac{X}{\Delta t}} \int_0^{\Delta t} \lambda_i(t)dt \right] + C_{pm} \times N_5 \qquad [3.15]$$

– *Policy II*:

- *Policy II-1*

$$\xi_{MII-1y} - \xi_{MX} \leq \xi_{MII-1n} \rightarrow \xi_{MX} \geq \xi_{MII-1y} - \xi_{MII-1n} \rightarrow \xi_{MX} \geq B_{M1}$$

$$\xi_{MII-1n} = C_{cm} \times \left[\sum_{j=0}^{N_1} \sum_{i=In\left(j\cdot\frac{T}{\Delta t}\right)+1}^{In\left((j+1)\cdot\frac{T}{\Delta t}\right)} \int_0^{\Delta t} \lambda_i(t)dt + \sum_{i=In\left(N_1\cdot\frac{T}{\Delta t}\right)+1}^{In(\frac{X}{\Delta t})} \int_0^{\Delta t} \lambda_i(t)dt \right] + C_{pm} \times N_1$$

$$\xi_{MII-1y} = C_{cm} \times \left[\sum_{j=0}^{N_1} \sum_{i=In\left(j \cdot \frac{T}{\Delta t}\right)+1}^{In\left((j+1) \cdot \frac{T}{\Delta t}\right)} \int_0^{\Delta t} \lambda_i(t)dt + \sum_{i=In\left(N_1 \cdot \frac{T}{\Delta t}\right)+1}^{In\left(\frac{X}{\Delta t}\right)} \int_0^{\Delta t} \lambda_i(t)dt + \sum_{i=In\left(\frac{X}{\Delta t}\right)}^{In\left(\frac{X_e}{\Delta t}\right)} \int_0^{\Delta t} \lambda_i(t)dt \right] + C_{pm} \times N_1$$

$$B_{M1} = C_{cm} \times \left[\sum_{i=In\left(\frac{X}{\Delta t}\right)}^{In\left(\frac{X_e}{\Delta t}\right)} \int_0^{\Delta t} \lambda_i(t)dt \right] \qquad [3.16]$$

– *Policy II-2*:

$$\xi_{MII-2y} - \xi_{MX} \le \xi_{MII-2n} \rightarrow \xi_{MX} \ge \xi_{MII-2y} - \xi_{MII-2n} \rightarrow \xi_{Mw} \ge B_{M2}$$

$$\xi_{MII-2n} = C_{cm} \times \left[\sum_{j=0}^{N_1} \sum_{i=In\left(j \cdot \frac{T}{\Delta t}\right)+1}^{In\left((j+1) \cdot \frac{T}{\Delta t}\right)} \int_0^{\Delta t} \lambda_i(t)dt + \sum_{i=In\left(N_1 \cdot \frac{T}{\Delta t}\right)+1}^{In\left(\frac{X}{\Delta t}\right)} \int_0^{\Delta t} \lambda_i(t)dt \right] + C_{pm} \times N_1$$

$$\xi_{MII-2y} = C_{cm} \times \left[\sum_{j=0}^{N_4} \sum_{i=In\left(j \cdot \frac{T}{\Delta t}\right)+1}^{In\left((j+1) \cdot \frac{T}{\Delta t}\right)} \int_0^{\Delta t} \lambda_i(t)dt + \sum_{i=In\left(N_4 \cdot \frac{T}{\Delta t}\right)+1}^{In\left(\frac{X_e}{\Delta t}\right)} \int_0^{\Delta t} \lambda_i(t)dt \right] + C_{pm} \times N_4$$

$$B_{M2} = C_{cm} \times \left[\sum_{j=0}^{N_4} \sum_{i=In\left(j \cdot \frac{T}{\Delta t}\right)+1}^{In\left((j+1) \cdot \frac{T}{\Delta t}\right)} \int_0^{\Delta t} \lambda_i(t)dt + \sum_{i=In\left(N_4 \cdot \frac{T}{\Delta t}\right)+1}^{In\left(\frac{X_e}{\Delta t}\right)} \int_0^{\Delta t} \lambda_i(t)dt - \sum_{j=0}^{N_1} \sum_{i=In\left(j \cdot \frac{T}{\Delta t}\right)+1}^{In\left((j+1) \cdot \frac{T}{\Delta t}\right)} \int_0^{\Delta t} \lambda_i(t)dt - \sum_{i=In\left(N_1 \cdot \frac{T}{\Delta t}\right)+1}^{In\left(\frac{X}{\Delta t}\right)} \int_0^{\Delta t} \lambda_i(t)dt \right] + C_{pm} \times (N_4 - N_1)$$

[3.17]

3.3.5. *Win–win interval for the extended warranty cost*

There are instances when there is a win–win interval with regards to the extended warranty cost taking into account the maximum additional cost that the lessee must pay and the minimum price at which the lessor can sell the extended warranty. We will now determine a theoretical condition under which a win–win interval exists where the maximum additional cost for the lessee is greater than the minimum selling price for the lessor.

– *Policy I-1*: using equations [3.12] and [3.15], there is an area of financial compromise for the extended warranty where:

$$
\begin{aligned}
A_{M1} \leq A_{cI} \Rightarrow C_{cm} \times \Bigg[& \sum_{i=1}^{In\left(\frac{X}{\Delta t}\right)} \int_0^{\Delta t} \lambda_i(t)dt \\
& + \sum_{j=0}^{N_3-1} \sum_{i=In\left(j\cdot\frac{T}{\Delta t}\right)+\frac{X}{\Delta t}+1}^{In\left((j+1)\cdot\frac{T}{\Delta t}\right)+\frac{X}{\Delta t}} \int_0^{\Delta t} \lambda_i(t)dt + \sum_{i=In\left(N_3\cdot\frac{T}{\Delta t}\right)+\frac{X}{\Delta t}+1}^{In\left(\frac{L}{\Delta t}\right)} \int_0^{\Delta t} \lambda_i(t)dt \\
& - \sum_{i=In\left(N_5\cdot\frac{T}{\Delta t}\right)+\frac{X}{\Delta t}+1}^{In\left(\frac{X_e}{\Delta t}\right)} \int_0^{\Delta t} \lambda_i(t)dt \\
& - \sum_{j=0}^{N_6-1} \sum_{i=In\left(j\cdot\frac{T}{\Delta t}\right)+\frac{X_e}{\Delta t}+1}^{In\left((j+1)\cdot\frac{T}{\Delta t}\right)+\frac{X_e}{\Delta t}} \int_0^{\Delta t} \lambda_i(t)dt - \sum_{i=In\left(N_6\cdot\frac{T}{\Delta t}\right)+\frac{X_e}{\Delta t}+1}^{In\left(\frac{L}{\Delta t}\right)} \int_0^{\Delta t} \lambda_i(t)dt \\
& - \sum_{j=0}^{N_5-1} \sum_{i=In\left(j\cdot\frac{T}{\Delta t}\right)+\frac{X}{\Delta t}+1}^{In\left((j+1)\cdot\frac{T}{\Delta t}\right)+\frac{X}{\Delta t}} \int_0^{\Delta t} \lambda_i(t)dt \Bigg] + C_{pm} \times (N_3 - N_6 - N_5) \geq 0
\end{aligned}
$$

[3.18]

– *Policy II-1:* using equations [3.13] and [3.16], there is an area of financial compromise for the extended warranty where:

$$B_{c1} \geq B_{M1} \Rightarrow C_{cm} \times \left[\sum_{j=0}^{N_3-1} \sum_{i=In\left(j\cdot\frac{T}{\Delta t}\right)+\frac{X}{\Delta t}+1}^{In\left((j+1)\cdot\frac{T}{\Delta t}\right)+\frac{X}{\Delta t}} \int_0^{\Delta t} \lambda_i(t)dt \right.$$

$$+ \sum_{i=In\left(N_3\cdot\frac{T}{\Delta t}\right)+\frac{X}{\Delta t}+1}^{In\left(\frac{L}{\Delta t}\right)} \int_0^{\Delta t} \lambda_i(t)dt - 2 \times \sum_{i=In\left(\frac{X}{\Delta t}\right)}^{In\left(\frac{X_e}{\Delta t}\right)} \int_0^{\Delta t} \lambda_i(t)dt$$

$$\left. - \sum_{j=0}^{N_6-1} \sum_{i=In\left(j\cdot\frac{T}{\Delta t}\right)+\frac{X_e}{\Delta t}+1}^{In\left((j+1)\cdot\frac{T}{\Delta t}\right)+\frac{X_e}{\Delta t}} \int_0^{\Delta t} \lambda_i(t)dt - \sum_{i=In\left(N_6\cdot\frac{T}{\Delta t}\right)+\frac{X_e}{\Delta t}+1}^{In\left(\frac{L}{\Delta t}\right)} \int_0^{\Delta t} \lambda_i(t)dt \right]$$

$$+ C_{pm} \times (N_3 - N_6) > 0 \qquad [3.19]$$

– *Policy II-2*: using equations [3.14] and [3.17], there is an area of financial compromise for the extended warranty where:

$$B_{c2} \geq B_{M2} \Rightarrow C_{cm} \times \left[\sum_{i=In\left(N_1\cdot\frac{T}{\Delta t}\right)+1}^{In\left(\frac{X}{\Delta t}\right)} \int_0^{\Delta t} \lambda_i(t)dt \right.$$

$$+ \sum_{j=0}^{N_3-1} \sum_{i=In\left(j\cdot\frac{T}{\Delta t}\right)+\frac{X}{\Delta t}+1}^{In\left((j+1)\cdot\frac{T}{\Delta t}\right)+\frac{X}{\Delta t}} \int_0^{\Delta t} \lambda_i(t)dt + \sum_{i=In\left(N_3\cdot\frac{T}{\Delta t}\right)+\frac{X}{\Delta t}+1}^{In\left(\frac{L}{\Delta t}\right)} \int_0^{\Delta t} \lambda_i(t)dt$$

$$- \sum_{i=In\left(N_5\cdot\frac{T}{\Delta t}\right)+\frac{X}{\Delta t}+1}^{In\left(\frac{X_e}{\Delta t}\right)} \int_0^{\Delta t} \lambda_i(t)dt$$

$$- \sum_{j=0}^{N_6-1} \sum_{i=In\left(j\cdot\frac{T}{\Delta t}\right)+\frac{X_e}{\Delta t}+1}^{In\left((j+1)\cdot\frac{T}{\Delta t}\right)+\frac{X_e}{\Delta t}} \int_0^{\Delta t} \lambda_i(t)dt$$

$$- \sum_{i=In\left(N_6\cdot\frac{T}{\Delta t}\right)+\frac{X_e}{\Delta t}+1}^{In\left(\frac{L}{\Delta t}\right)} \int_0^{\Delta t} \lambda_i(t)dt - \sum_{j=0}^{N_4} \sum_{i=In\left(j\cdot\frac{T}{\Delta t}\right)+1}^{In\left((j+1)\cdot\frac{T}{\Delta t}\right)} \int_0^{\Delta t} \lambda_i(t)dt$$

$$- \sum_{i=In\left(N_4\cdot\frac{T}{\Delta t}\right)+1}^{In\left(\frac{X_e}{\Delta t}\right)} \int_0^{\Delta t} \lambda_i(t)dt + \sum_{j=0}^{N_1} \sum_{i=In\left(j\cdot\frac{T}{\Delta t}\right)+1}^{In\left((j+1)\cdot\frac{T}{\Delta t}\right)} \int_0^{\Delta t} \lambda_i(t)dt$$

$$\left. + \sum_{i=In\left(N_1\cdot\frac{T}{\Delta t}\right)+1}^{In\left(\frac{X}{\Delta t}\right)} \int_0^{\Delta t} \lambda_i(t)dt \right] + C_{pm} \times (N_3 - N_6 - N_4 + N_1) > 0 \qquad [3.20]$$

3.4. Numerical example

In order to illustrate the model we have developed in this chapter, we will consider a production/maintenance forecasting problem for a company that has leased a machine that has to satisfy a stochastic demand that is assumed to be Gaussian at a specified service level over a finite leasing horizon. The number L of leasing periods Δt is equal to 24, with Δt = 1 um. The leasing machine has a degradation law characterized by a Weibull distribution with shape parameter δ = 2 and scale parameter β = 100 (with these two parameters, the degradation is linear: γ = 2). From the failure rate equation, we will determine the average number of failures, assuming that after each preventive maintenance action the equipment is "as good as new".

The following arbitrarily chosen input data are also considered:

– Cpr1 = 3 mu;

– Cpr2 = 10 mu;

– service level: $\alpha = 0.95$;

– Cs = 5 mu;

– initial inventory: $S_0 = 20$;

– variation in demand: $V_{d_k} = 1.21$;

– $X = 2$;

– $Xe = 6$;

– Ccm = 1,500;

– Cpm = 200.

To compute the failure rate, we assume that the nominal degradation follows a Weibull distribution given by:

$$\lambda_n(t) = \frac{\gamma}{\beta} \cdot \left(\frac{t}{\beta} \right)^{\gamma - 1}$$

The average of forecasting demand is presented in Table 3.1.

d_1	d_2	d_3	d_4	d_5
15	**17**	**15**	**15**	**15**
d_6	d_7	d_8	d_9	d_{10}
14	***16***	**14**	**16**	**13**
d_{11}	d_{12}	d_{13}	d_{14}	d_{15}
15	**14**	**15**	**12**	**15**
d_{16}	d_{17}	d_{18}	d_{19}	d_{20}
13	***15***	***11***	***16***	***13***
d_{21}	d_{22}	d_{23}	d_{24}	d_{25}
15	***12***	***14***	***16***	***14***

Table 3.1. *Average forecasted demand*

Applying our analytical model, we used the numerical algorithms for constrained global optimization (Nelder–Mead) methods with the Mathematica program in order to realize this optimization. First, we want to find the optimal production plan, which is presented in Table 3.2. According to the production plan obtained, we observe, for each maintenance policy, the optimal preventive maintenance interval T^* and the existence of a win–win interval where the lower and upper boundaries are, respectively, the minimum price at which the lessor should sell the extended warranty, and the maximum additional cost that the lessee should pay for the extended warranty.

$u^*(1)$	$u^*(2)$	$u^*(3)$	$u^*(4)$	$u^*(5)$
9	**14**	**8**	**12**	**12**
$u^*(6)$	$u^*(7)$	$u^*(8)$	$u^*(9)$	$u^*(10)$
15	9	**13**	**14**	**11**
$u^*(11)$	$u^*(12)$	$u^*(13)$	$u^*(14)$	$u^*(15)$
10	**5**	**11**	**12**	**5**
$u^*(16)$	$u^*(17)$	$u^*(18)$	$u^*(19)$	$u^*(20)$
15	**16**	**12**	**10**	**6**
$u^*(21)$	$u^*(22)$	$u^*(23)$	$u^*(24)$	$u^*(25)$
2	**5**	**17**	**3**	**14**

Table 3.2. *Plan to meet the forecast production requirements*

From Table 3.3 we can determine, for example, that for Policy II.2 (preventive maintenance performed during [0, *Xe*)) the optimal preventive maintenance interval for the lessor is equal to $T^*_M = 3$, and for the lessee it is equal to $T^*_C = 2$. The win–win interval for the extended warranty cost is between 202.25 and 203.476 mu (Figure 3.5), which are the threshold values for the lessor and the lessee, respectively. In this case, the best compromise corresponds to the value in the middle of this interval, which works out at an extended warranty cost of 202.863 mu.

T	Policy I-1		Policy II-1		Policy II-2	
	Lessor	**Lessee**	**Lessor**	**Lessee**	**Lessor**	**Lessee**
1	803,724	804,288	3.41471	800.309	803.194	803.194
2	403,72	404,288	3.41471	400.309	403.194	203.476
3	**202,347**∅	**194,706**	**3.41471**	**200.238**	**202.25**	**403.759**
4	203,724	204,288	3.41471	200.309	203.759	403.194

Table 3.3. *Compromise intervals for the extended warranty cost (in mu) for Cpm = 200 and Ccm = 1,500*

In fact, from the lessor's perspective, as preventive maintenance actions become more efficient, the average number of failures decreases. Consequently, he/she would pay less for minimal repairs and therefore his/her threshold value for the extended warranty cost becomes lower. From the lessor's perspective, taking the extended warranty will result in the leasing machine entering the post-warranty period with higher reliability, thanks to preventive maintenance actions performed during $[X, Xe]$. Consequently, the lessee is willing to pay more for the extended warranty. Thus, preventive maintenance becomes more efficient, giving a higher reliability and minimal average number of failures, and hence the lowest number of minimal repairs (with attendant repair costs) during the post-warranty period.

For Policy II.1, since the period during which preventive maintenance is performed is only related to the basic warranty period, the win–win interval for the extended warranty cost is between 3.41471 mu and 200.238 mu. Since there is no preventive maintenance for the lessor, the minimal expected number of repairs during this period remains the same with or without the extended warranty; whereas for the lessee, the optimal preventive maintenance interval is equal to $T^*_c = 3$.

From Figure 3.6, we can see for Policy I-1 that choosing an extended warranty period would not be of interest to the lessee nor for the lessor. There is no win–win interval because the minimum price at which the lessor should sell the extended warranty (202.347 mu) is greater than the maximum additional cost that the lessee should pay for the extended warranty (194,706 mu). The extended warranty would not be as advantageous for the lessor as for the lessee due to the fact that there is no preventive maintenance, and hence the average number of failures during [X, Xe] remains the same with or without the extended warranty.

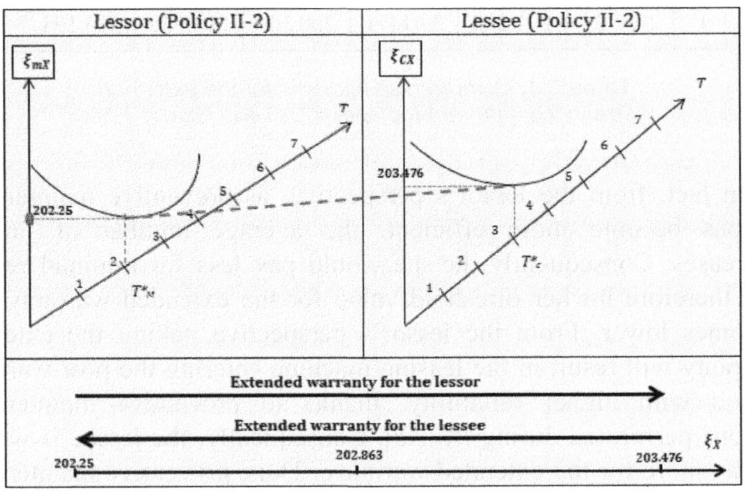

Figure 3.5. *The win–win interval for the extended warranty for Policy II-2*

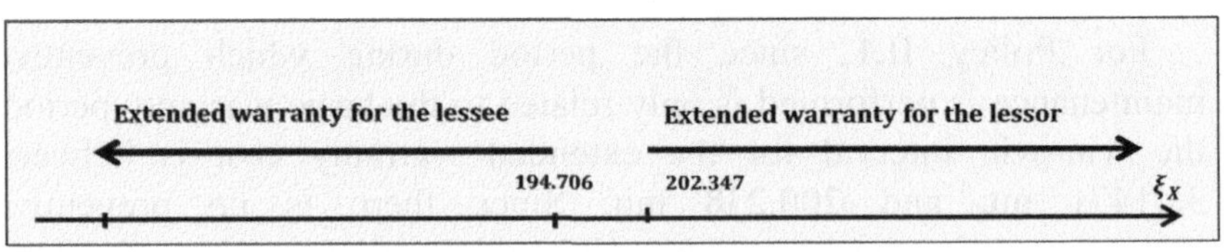

Figure 3.6. *There is no win–win interval for the extended warranty cost for Policy I.1*

3.4.1. *Variation in preventive maintenance and corrective maintenance costs*

We will now identify the impact of varying the maintenance preventive cost (*Cpm*) and the maintenance corrective cost (*Ccm*) during the leasing horizon. Besides the nominal values (*Cpm* = 200 and *Ccm* = 1,500), we consider higher values ($Cpm \in \{300,400,600\}$; $Ccm \in \{1700,1900,2000\}$). The effect of varying the maintenance costs can be observed for different policies. For different policies, we can see that the period over which preventive maintenance is performed has a direct impact on the average number of failures and on the minimum number of repairs during the extended warranty period [X, Xe].

From Tables 3.4 and 3.5 we can see that for Policy I.1 (where preventive maintenance actions are performed during [Xe, L)), the preventive maintenance interval is increased if the preventive and corrective costs are increased but it is not beneficial for either the lessee nor the lessor to adopt the extended warranty period whatever the maintenance costs. This can be explained by the fact that, at the end of warranty period *Xe*, it will be too late to start preventive maintenance action and that as the degradation of the leasing machine increases preventive maintenance actions will not improve the reliability of leasing machine, even if actions are performed more frequently during [Xe,L].

For Policies II.1 (preventive maintenance performed during [0,X)) and II.2 (preventive maintenance carried out during [0,Xe)), there is a different trend compared to Policy I.1, since win–win intervals exist for all values of *Cpm* and *Ccm*. These intervals become larger as the maintenance costs increase, but for Policy II.1 the minimum price at which the lessor should sell the extended warranty is fixed for all *Cpm* and *Ccm* values since there is no preventive maintenance for the lessor to carry out and the expected minimum number of repairs during this period remains the same with or without the extended warranty.

Cpm	Policy I-1		Policy II-1		Policy II-2	
	Lessor	Lessee	Lessor	Lessee	Lessor	Lessee
200	202,347	Ø 194,706	3.41471	200.238	202.25	203.476
300	302,34	Ø294,706	3,41471	300,238	300,282	303,759
400	402,34	Ø394,706	3,41471	400,238	400,282	403,759
600	602,347	Ø594,706	3,41471	600,238	600,282	603,759

Table 3.4. *Variation in Cpm cost when Ccm = 1,500*

Ccm	Policy I-1		Policy II-1		Policy II-2	
	Lessor	Lessee	Lessor	Lessee	Lessor	Lessee
1500	202.347	Ø 192.915	4.55294	200.238	200.282	203.759
1700	202.66	Ø191.97	4.55294	200.27	200.32	204.26
1900	202,973	Ø191,025	4.55294	200.302	200.358	204.761
2000	203.129	Ø 190,553	4.55294	200.318	200.376	205.012

Table 3.5. *Variation in Ccm cost when Cpm = 200*

3.4.2. *Effects of variation in production period length Δt*

In this section, we investigate the effects of varying the length (Δt) of production during the product's lifecycle. Beside the nominal value ($\Delta t = 1$), we consider a higher value ($\Delta t = 2$).

T	Policy I-1		Policy II-1		Policy II-2	
	Lessor	Lessee	Lessor	Lessee	Lessor	Lessee
1	903,830	904,180	3.41471	800.309	803.194	803.194
2	303,27	Ø305,28	4.7141	500.39	504.194	504.76
3	345,347	304,706	4.91471	501.238	522.25	503.759
4	360,724	306,288	4.94417	511.309	533.759	513.140

Table 3.6. *Compromise intervals for the extended warranty cost at Cpm = 200 and Ccm = 1,500 and Δt = 2*

The effect of varying the production period can only be observed for Policy II.1 and Policy II.2. The optimal preventive maintenance interval for the lessor is decreased for production period $\Delta t = 2$ (Policy II.1: $T^*_M = 2$ and $T^*_C = 2$) (Table 3.6) with a higher cost relative to $\Delta t = 1$ (Policy II.1: $T^*_M = 3$ and $T^*_C = 3$) (Table 3.3). The compromise interval for the extended warranty cost gets larger as the number of preventive maintenance actions increases (with increased production period). In fact, if the length of the production period or the demand increases, the principal machine produces more to meet customers' demands; thus the machine will undergo more failures and the preventive maintenance interval will increase. According to the previous results presented through the variability in Δt, the production period length is really impacted visibly.

For $Cpm = 200$ and $Ccm = 1{,}500$ and $\Delta t = 2$, the compromise intervals for the extended warranty cost are given in Table 3.6.

3.5. Conclusion

This chapter treats a forecasting production/maintenance problem correlated to the adoption of an extended warranty period for a leasing machine over a finite leasing horizon. First, we developed a mathematical model for a forecasting problem in order to determine a forecasting production plan. Second, an analytical model was proposed in order to study the opportunity provided by the extended warranty in a leasing contract from both the lessee and the lessor perspectives. We proposed different maintenance policies during the finite leasing horizon, which we have considered to be:

– the influence of production rates on the degree of degradation of the leasing machine;

– periodic preventive maintenance actions with different costs.

For each maintenance policy, we expressed the total cost incurred by the lessee and by the lessor in order to determine the maximum additional cost the lessee should pay for the extended warranty, and the minimum price at which the lessor should sell it. For each policy and for any given situation, conditions where a win–win interval exists between the lessee and the lessor have been found.

For future research, we will consider a more complex system with other types of warranty policies (including the number of warranty dimensions, the renewability of a warranty and warranty compensation methods). For the maintenance strategy, we will consider new hypotheses where the corrective and preventive times are not negligible.

4

Global Control Policy Taking into Account Maintenance and Product Non-conformity

This chapter presents the control policy of a manufacturing system consisting of two machines and two buffers. This production system generates conforming and non-conforming products. The control variables considered are the instants at which preventive maintenance has to be performed on each machine, and both buffer inventory levels. Our objective is to reach the best compromise with regard to cost, availability and quality. The approach adopted is based on simulation, experimental design and a multi-criteria analysis.

4.1. Introduction

Having a reliable production tool that makes conforming parts, respects deadlines and produces at low cost is a constant concern for manufacturing companies. Production strategies based on the "just-in-time" concept contribute to reaching the objectives of quality, time limits and costs. Given the hazards of breakdowns and resource availability, and in order to meet demand, it is often necessary to establish a preventive maintenance policy and have buffer stocks between the different machines. This must be done in order to have the best possible availability of the production tool at the lowest cost.

Several maintenance policies are suggested in the literature. The most popular basic strategies are the age-type and the block-type policies [BAR 65]. The mathematical models that govern them are meant to maximize the availability of the equipment and to minimize the average total cost of maintenance actions over an infinite horizon. Many extensions have been provided for these models [SHE 81, VAL 89, CHO 91, KEC 95].

The development of maintenance strategies for production lines made up of one or more machines has been the subject of several studies, such as those by Gershwin [GER 94] and Glassey *et al.* [GLA 93]. Some deal with predictive maintenance and imply equipment inspection [CHE 98, BAD 02] and others focus on equipment submitted to random shocks [PIE 76, SHE 95]. Wang [WAN 02] has presented an overview of the topic.

It is important to note that the majority of maintenance models only take into account data that are specific to maintenance, such as hazard functions associated with failures and repairs, as well as the costs of maintenance actions and spare parts. However there are an increasing number of studies based on a more global approach which jointly consider the parameters of maintenance and those relating to production, such as the size of the batches that must be produced, the size of buffer stocks, etc. For examples, readers can consult Buzacott *et al.* [BUZ 92], Dallery *et al.* [DAL 92], Xie [XIE 93], Van Bracht [VAN 95b] and Chelbi *et al.* [CHE 04] on the assessment of production systems subject to random breakdowns. Chan [CHA 01] suggests a simulation model to assess the performance of a production line operating in "push production". Tempelmeier [TEM 01] considers the assessment of the performance of a non-homogeneous production line depending on the maintenance parameters of the machines and the quality of the products made. Chen *et al.* [CHE 97] considers optimizing the buffer stock level and taking into account the age of the machine on which preventive maintenance must be performed.

Other studies focus on the control of production systems within which preventive maintenance is carried out [GHA 00, KEN 01].

These studies develop analytical approaches supplemented with simulations.

From a general point of view, analysis aiming to optimize production systems made up of several machines and producing a variety of products gives rise in the literature to Markovian-type mathematical models, which lead to the Hamilton–Jacobi–Bellman equations whose rigorous analytical resolution is tricky, if not impossible, except in relatively simple cases [AKE 86]. Solving such a problem generally involves breaking down the system into subsets (composed of a machine and buffer stock) and determining sub-optimality parameters. Several methodologies are presented in the literature that enable us to determine sub-optimality parameters, for example the double-threshold control [VAN 93], the hierarchical heuristic approach [BAI 95] and the hierarchical asymptotic method [SET 94]. Let us also note the methodology suggested by Burman [BUR 95], and that proposed by Dallery *et al.* [DAL 99], which are based on a resolution algorithm in the case of non-homogeneous transfer lines.

Another resolution method suggested by Gharbi *et al.* [GHA 03] is based on an analytical approach coupled with simulation and experimental design. The authors suppose that all the items produced conform (i.e. there is no rejection), that the breakdown and repair rates of machines are constant, and that there are no preventive maintenance actions. In their models they view the production rates as control variables.

In this chapter we consider a production system made up of two machines and two buffer stocks (intermediary and final) making a unique type of product where products that do not conform are rejected at a specified rate. The failure rate of each machine increases over time. The machines are submitted to an age-type preventive maintenance policy that aims to improve machine availability and reduce the production of non-conforming parts, which directly affect the stock levels and cause losses.

The objective of this chapter is to determine an integrated maintenance-control policy. It has been proven in previous studies

[OUA 02, REZ 04] that an integrated maintenance-control strategy is more economical than a basic maintenance strategy separating control (stock management) and maintenance. We intend to simultaneously determine the preventive maintenance periods on each machine as well as stock levels in order to reach the best compromise between cost, availability and quality. To solve this problem we use simulation, experimental design and a multi-criteria analysis method taking into account the cost criterion, the stationary availability of the system (which should be maximized) and that of the number of parts rejected (which should be minimized).

Section 4.2 of this chapter reviews the basic analytical model, which describes the control of multi-machine and multi-product stochastic production systems. Considering the difficulty of finding an optimal solution through this analytical approach and the restrictive hypotheses it implies compared to the situation of the system analyzed (section 4.3), an approach based on simulation and experimental design will be taken in sections 4.4 and 4.5 in order to analyze this system. A multi-criteria approach is presented in section 4.6 in order to reach the best compromise between cost, availability and quality. Finally, the last section presents the conclusions of this study and discusses some future prospects.

4.2. Control strategy for stochastic multi-machine multi-product systems: analytical approach

Let us consider a production system made up of m machines subject to random breakdowns and producing n types of products P_j ($j = 1, 2, ..., n$). The breakdown and repair rates of the m machines are assumed to be constant. Adjustment time and costs are considered to be negligible. The mathematical model and the optimality conditions corresponding to this system are presented in detail in Gharbi *et al.* [GHA 03]. Sections 4.2.1 and 4.2.2 give the notations used and the formula for the updated average total cost, which represents the functions that need to be minimized.

4.2.1. *Notations*

x_i	level of the inventory associated with machine M_i ($i = 1,\ldots, m$)
$X(t)$	vector level of inventories; $(x_1,x_2,\ldots,x_m)$ = state of the system vector
u_{ij}	production rate of product P_i on machine M_i
$U(t)$	vector production rate; $(u_1,u_2,\ldots u_m)$ = control vector
d	vector representing demand
$g(\cdot)$	immediate cost function
	$x_j^+(\cdot)=\max(0,x_j(\cdot))$ and $x_j^-(\cdot)=\max(-x_j(\cdot),0)$
c_j^+ and c_j^-	unit costs (per time unit) of storage and shortage (respectively) of product P_j with $j = 1, \ldots, n$
ρ	discount rate
$E[A/B]$	probability of occurrence of A, B having occurred
$\xi_i(t)$	state of machine i at instant t ($\xi_i(t)$ = 1 if the machine is available, $\xi_i(t)$ = 2 if it is not)
	$\text{Ind}\{\xi_i(t)=\delta\}=\begin{cases}1 \text{ if } & \xi_i(t)=\delta \\ 0 \; otherwise\end{cases} \quad i = 1,\ldots,m$
$q_{12}^{\,i}$	transition rate from state 1 to state 2 of machine M_i
$J(\cdot)$	updated average total cost function

4.2.2. *Formulation of the cost optimization problem*

The mathematical model presented in Gharbi *et al.* [GAR 03] leads to the following expression for the updated average total cost:

$$J(x,\alpha,u(.)=E\left[\int_0^\infty e^{-\rho t}g(x(t),u(t))dt) \quad | \quad x(t)=x, \quad \xi(t)=\alpha\right] \qquad [4.1]$$

The problem consists of determining an acceptable solution u(˙) ∈ U(˙) that minimizes function J(˙) while taking into account the following constraints:

– the stock variation:

$$\overset{\circ}{x}(t) = u(t) - d \quad [4.2]$$

– the constraints related to production control:

$$u_{ik} = 0 \quad \forall k \neq j, \quad i = 1,\ldots,m \quad [4.3]$$

– the characteristics of the system in order to meet downstream demand:

$$d_j < \frac{1}{nm} \sum_{i=1}^{m} u_{\max}^{ij} \sum_{i=1}^{m} \frac{q_{12}^{i}}{q_{12}^{i} + q_{21}^{i}} \quad [4.4]$$

– the value of the immediate cost function:

$$g(x(\cdot),\cdot) = \sum_{j=1}^{n} (c_j^{+} x_j^{+}(\cdot) + c_j^{-} x_j^{-}(\cdot)) \quad [4.5]$$

The problem can be formulated by means of the following function ν(.,α), which considers the initial mode of machine availability α:

$$\upsilon(x,\alpha) = \inf_{u \in U(\alpha)} J(x,\alpha,u). \quad [4.6]$$

It has been proven in Akella *et al.* [AKE 86] that under certain hypotheses, function ν(x,α) is the solution to Hamilton–Jacobi–Bellman equations:

$$\rho \upsilon(x,\alpha) = \min_{u \in U(\alpha)} \{(u-d) \times \upsilon_x(x,\alpha) + Q\upsilon(x,\cdot)(\alpha) + g(x,\alpha)\} + g(x,u) \quad \forall x \in R^n \quad [4.7]$$

where $\nu_x(x,\alpha)$ represents the partial derivative of function υ(x,α). This function is convex and the control policy associated with it is optimal.

4.2.3. *Complexity of the optimal control problem*

The size of the Hamilton–Jacobi–Bellman equations is given by:

$$Dim = 2^m \times 3^{mxn} \times \prod_{j=1}^{n} N_h(x_j)$$

where $N_h(x_j)$ represents the card $[G_h(x_j)]$, and $G_h(x_j)$ represents a digital grid giving the relation between variables x_j` and products P_j, $j = 1,\ldots,n$.

For example, Gharbi *et al.* [GHA 03] show that for a system made up of two machines and producing five products (m = 2, n = 5) and $N_h(x_j)$ = 100, j = 1, ..., 5, the dimension equation gives $2.36*10^{15}$ states. The extreme complexity of creating a resolution algorithm for such a system is obvious.

Moreover, similarly to the production system considered in this chapter, which will be presented in section 4.3, with non-constant breakdown and repair rates (i.e. with non-Markovian processes) and with non-zero rejection rates it is not possible to establish optimality conditions like those given by equation [4.7]. No optimal control policy is available in such cases.

4.3. Description of the production system and the control strategy

The considered system (Figure 4.1) is composed of two machines M_i (i = 1, 2) in series, making only one type of product to meet demand for an assembly line, through a final stock S_2. Demand d is characterized by an arrival frequency and the quantity demanded. The frequency and the quantity can be random; they are represented by their respective means. For the example discussed in this chapter, periodicity and demand are ten time units and four parts, respectively. The two machines M_i deteriorate when used and are maintained through an age-type preventive maintenance program, i.e. for each machine a preventive maintenance is performed after m_i time units without any breakdown and the machines are refurbished after each breakdown.

A buffer stock S_1 is placed between the two machines in order to reduce the risk of not meeting demand in the situation when machine M_1 is unavailable after a breakdown.

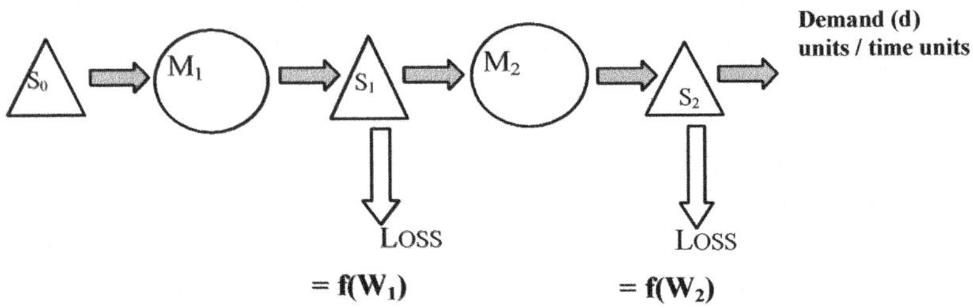

Figure 4.1. *The considered production system*

The two machines can operate at a production rate $\alpha_{i\ max}$ that is superior to upstream demand and so it permits S_i stocks to be rebuilt. The stock management policy consists of producing stock at maximum speed until the maximum level of stock, S_i, is reached; at this point the production level is reduced in order to simply meet demand.

It is considered that each machine may produce non-conforming parts. The number of non-conforming parts is proportional to the operating time, W_i [LEE 85, ROS 86, DOH 98]. These parts are removed from the production line in S_i stocks. Our goal is to determine the operational characteristics of the production system (age of each machine when it needs to undergo preventive maintenance and the levels of both stocks) in order to reach the best compromise between cost, availability and quality.

The following working hypotheses are considered:

– the probability distribution functions associated with the operating lives of the two machines and repair times are known;

– breakdowns are detected instantaneously;

– maintenance actions are performed perfectly and make the machine "as good as new";

– the demands that cannot be met through stock S_2 are lost;

– all the costs relating to maintenance and stock management are known and constant;

– all the resources required to perform maintenance actions are always available at the right time.

As shown in section 4.2, the analytical treatment of this problem is very complex, if not impossible. In the next two sections an approach based on simulation and experimental design is taken in order to analyze the production system. The goal is to obtain relatively simple mathematical functions for total cost per time unit and for stationary availability corresponding to a given configuration of input parameters that describe the system.

4.4. Simulation model

4.4.1. *Simulation principle*

The simulation principle used is based on potential events that may occur from a given state (see Table 4.1). Let us consider the subset made up of machine M_1 and its upstream stock. From the different states i (i = 1 ...4) of this subset, the graph of events e_{ij} and e'_{ij} that may occur is determined (see Figure 4.2):

– $j = d_{mp}$: beginning of preventive maintenance;

– $j = f_{mc}$: end of corrective maintenance;

– $j = d_{mc}$: beginning of corrective maintenance;

– etc....

State	Description	Possible events					
		Beginning of Mp (dmp)	End of Mp (fmc)	Beginning of Mc (dmc)	End of Mc (fmc)	Upstream demand (d)	Part production (a)
1	M_1 in operation	e_{1dmp}		e_{1dmc}		e_{1d}	$e_{1a} + e'_{1a}$
2	M_1 in preventive maintenance		$e_{2fmp} + e'_{2fmp}$			e_{2d}	
3	M_1 in corrective maintenance	e_{3dmp}			e_{3fmc}	e_{3d}	
4	M_1 stopped Stock overload	e_{4dmp}				e_{4d}	

Table 4.1. *The correspondence between states and events*

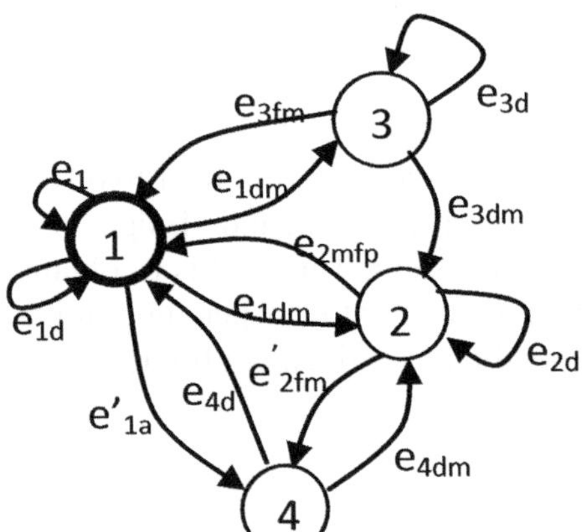

Figure 4.2. *The different states of the system*

4.4.2. *Simulation algorithm*

The following simulation algorithm has been created and programmed using the ProModel commercial software package.

The notations used are:

– $C_{CUM_}$i: cumulative cost for machine I;

– $\Gamma^{i}_{AP}(h_i, m_i)$: cost of stock management with and without shortage;

– T_{REP_i} cumulative maintenance time for machine i.

The inputs are:

– m_1: maintenance periodicity for machine M_1;

– m_2: maintenance periodicity for machine M_2;

– h_1: value of stock S_1;

– h_2: value of stock S_2;

– α_1: production capacity of machine M_1;

– α_2: production capacity of machine M_2;

– α_3: upstream demand *d*;

– M_p: average cost of a preventive maintenance action;

– M_c: average cost of a corrective maintenance action;

– T_{SIMUL}: simulation time.

It may be recalled that m_1, m_2, h_1 and h_2 represent the decision variables.

The outputs are:

– CT: average total cost per time unit;

– DISP: stationary availability of the whole system;

– Cumul_loss: cumulative loss (non-conforming products).

For each i, (i = 1, 2) :

1) Generate the breakdown time W_i.

2) ***If*** $W_i \geq m_i$,

then $C_{CUM}\_i = C_{CUM}\_i + M_p$

otherwise $C_{CUM}\_i = C_{CUM}\_i + M_c$.

3) Determine loss = $f(\min(W_i, m_i))$

Cumul_loss = Cumul_loss + loss.

4) Generate a repair time D_i

$T_{REP\_i} = T_{REP\_i} + D_i$

$T_{running\_i} = T_{running\_i} + \min(W_i, m_i) + D_i$.

5) ***If*** $D_i > \frac{h_i}{\alpha_{i+1}}$

then $C_{CUM} = C_{CUM} +$

otherwise $C_{CUM\_i} = C_{CUM\_i} + \Gamma^{i}_{SP}(h_i, m_i)$.

6) ***If*** $T_{running\_i} < T_{SIMUL}$

then back to stage 1.

7) $CT_i = C_{CUM\_i} / T_{running\_i}$

$$CT = \sum_{i=1}^{2} CT_i$$

$$DISP = (T_{running\_i} - \Sigma T_{REP\_i}) / T_{running\_i}.$$

8) END.

The following numerical data, which are arbitrarily selected, will be used in the rest of the chapter to illustrate the adopted approach:

– the density function associated with the operating life of each machine is a Weibull law with a shape of $\alpha = 2$ and a scale of $\beta = 100$;

– the durations of corrective and preventive maintenance actions follow an exponential law whose means are 20 and 10 time units, respectively;

– $M_{p\_1} = M_{p\_2} = 300$ mu (monetary units), $M_{c\_1} = M_{c\_2} = 2{,}000$ mu;

– storage cost C_{s_i} is 4 mu per time unit for the products in stocks S_1 and S_2;

– shortage cost C_{p_i} is 260 mu for each unit lost;

– the upstream demand *d* to be met is defined by an average quantity of four parts and a frequency of occurrence of 10 time units.

Once these data have been selected, the total cost per time unit, as well as stationary availability, will only depend on the independent variables h_1, h_2, m_1 and m_2. In the following section we suggest a methodology based on experimental design and variance analysis and on the notion of desirability in order to determine the best compromise between cost availability and quality

4.5. Experimental analysis

4.5.1. *Principle of the analysis*

Mathematical modeling of the cost function or of the availability function consists of finding, from a series of tests, a function $\Pi = f(X_i)$ where X_i are the decision variables. In this case the variables are m_1, h_1, m_2 and h_2.

In the first simulation enabling us to exclude a first-order mathematical model, we are trying to represent each function (cost and availability) by means of a quadratic model of the type:

$$\Pi = \alpha_0 + \sum_{i=1}^{k} \alpha_i.X_i + \sum_{i=1}^{k-1} \sum_{j=2.(j>i)}^{k} \alpha_{ij}.X_i.X_j + \sum_{i=1}^{i=k} \alpha_{ii}.X_i^2$$

with α_i and α_{ij} representing coefficients to be determined.

A second series of off-line simulations has enabled us to select intervals for variables m_1, h_1, m_2 and h_2. We have:

- $m_1 \in [10, 390]$;
- $m_2 \in [10, 390]$;
- $h_1 \in [1, 49]$;
- $h_2 \in [1, 49]$.

Three-level standardized variables have been used: –1, 0 and 1. The relations between the real variables (m_i, h_i) and the standardized variables (X_{mi}, X_{hi}) are given by the data in Table 4.2.

	Level –1	Level 0	Level 1
m_i	10	200	390
h_i	1	25	49

Table 4.2. *Relations between real variables and standardized variables*

The different coefficients α_i and α_{ij} were obtained through multi-linear regression on the 81 experimental tests carried out. Full factorial designs with four factors and three levels were used in order to establish a second-order model.

The simulations were performed over 5,000 time units. The results were derived from the mean of five replications.

4.5.2. *Determination and validation of the cost function*

The regression analysis method [SAD 91] and the use of the STATGRAPHICS software package serve as the basis of this model when the value of each coefficient of the cost function (Π_1), described underneath, has been estimated.

The cost function associated with the data used is:

$$\Pi_1 = 23.0973 + 0.179808*X_{m1} + 0.113333*X_{h1} - 1.52167*X_{m2} - 3.35722*X_{h2} + 0.829423*X_{m1}^2 + 0.35625*X_{m1}*X_{h1} - 0.16875*X_{m1}*X_{m2} + 0.32875*X_{m1}*X_{h2} + 0.391154*X_{h1}^2 + 0.26875*X_{h1}*X_{m2} + 1.17375*X_{h1}*X_{h2} + 0.966154*X_{m2}^2 + 2.03375*X_{m2}*X_{h2} + 2.76615*X_{h2}^2$$

Analysis of variance permits us to statistically test the effect of each factor and interaction on the total cost per time unit. To do so, for each coefficient we determine:

1) the squared sum of deviations, SCE, given by the formula:

$$SCE_K = \frac{N}{n_k} \sum_{i=1}^{i=n_i} (E_{Ki})^2$$

where:

– SCE_K is the squared sum of deviations of coefficient K;

– N is the total number of tests in the experimental design;

– n_K is the number of levels for factor K;

– E_{Ki} is the effect of factor K at level I;

2) the variance given by the relation with *ddl*(*K*), which is the degrees of freedom of factor *K*;

3) the percentage contribution of coefficient *i* given by relation

$$\frac{SCE_K}{SCE_{Tot}}$$

where SCE_{Tot} is the sum total of the squared deviations;

4) the ratio of variance of the factor (or of the interaction) to residual variance is denoted $F_{calculated}$;

5) the coefficient $F_{theoretical}$ is determined thanks to the Fisher-Snédécor table (5% risk, $v_1 = 14$, $v_2 = 12$; that is 3.13 in our example).

If the $F_{calculated}$ ratio is superior to $F_{theoretical}$, the test is considered statistically significant (S) and the effect of the factor is then highlighted. Otherwise it will be non-significant (NS). Table 4.3 shows the results obtained.

Source	SCE	ddl	Variance	% Cont.	$F_{calculated}$	Test
$\alpha_1 : m_1$	0.524	1	0.524	0.13	0.22	NS
$\alpha_2 : h_1$	0.231	1	0.231	0.06	0.10	NS
$\alpha_3 : m_2$	**41.678**	**1**	**41.678**	**10.71**	**17.44**	**S**
$\alpha_4 : h_2$	**202.877**	**1**	**202.877**	**52.15**	**84.89**	**S**
$\alpha_{11} : m_1^2$	1.077	1	1.077	0.28	0.45	NS
$\alpha_{12} : m_1.h_1$	2.030	1	2.030	0.52	0.85	NS
$\alpha_{13} : m_1.m_2$	0.456	1	0.456	0.12	0.19	NS
$\alpha_{14} : m_1.h_2$	1.729	1	1.729	0.44	0.72	NS
$\alpha_{22} : h_1^2$	0.356	1	0.356	0.09	0.15	NS
$\alpha_{23} : h_1.m_2$	1.156	1	1.156	0.30	0.48	NS
$\alpha_{24} : h_1.h_2$	**22.043**	**1**	**22.043**	**5.67**	**9.22**	**S**
$\alpha_{33} : m_2^2$	2.173	1	2.173	0.56	0.91	NS
$\alpha_{34} : m_2.h_2$	**66.178**	**1**	**66.178**	**17.01**	**27.69**	**S**
$\alpha_{44} : h_2^2$	**17.816**	**1**	**17.816**	**4.58**	**7.45**	**S**
Résidus	28.677	12	2.39	7.38		
TOTAL	389.001	26		100		

Table 4.3. *Results of the analysis of variance*

After the "pooling" operation (removing non-significant coefficients) the equation becomes:

$$\Pi_{10} = 23.0973 - 1.52167{*}X_{m2} - 3.35722{*}X_{h2} + 1.17375{*}X_{h1}{*}X_{h2} + 2.03375{*}X_{m2}{*}X_{h2} + 2.76615{*}X_{h2}^{2}$$

The optimal X_{m1}*, X_{h1}*, X_{m2}* and X_{h2}* are obtained by solving the following equations:

$$\left.\frac{\partial \Pi_1}{\partial X_{m1}}\right|_{X_{m1}=X_{m1}^*} = 0,$$

$$\left.\frac{\partial \Pi_1}{\partial X_{h1}}\right|_{X_{h1}=X_{h1}^*} = 0$$

and

$$\left.\frac{\partial \Pi_1}{\partial X_{h2}}\right|_{X_{h2}=X_{h2}^*} = 0$$

hence:

- $X_{m1}^* = -0.0402632$;
- $X_{h1}^* = -1$;
- $X_{m2}^* = 0.0974958$;
- $X_{h2}^* = 0.784789$;

which, returning to real values, gives us:

- $m_1^* =$ 197 time units;
- $h_1^* =$ 1 part;
- $m_2^* =$ 223 time units;
- $h_2^* =$ 43 parts;

for a minimum total cost of 21.54 mu.

Figures 4.3 and 4.4 show some iso-responses in the (m_2, h_2) plane and the cost function in the (h_1, m_2, h_2) space.

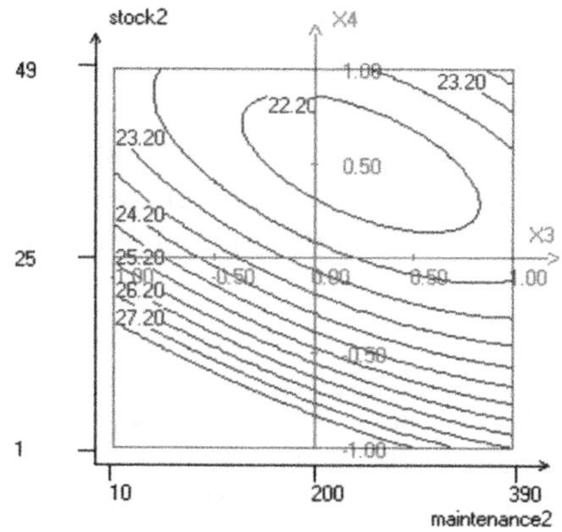

Figure 4.3. *Iso-response of the cost function in plane (m_2, h_2). For a color version of this figure, see www.iste.co.uk/rezg/services.zip*

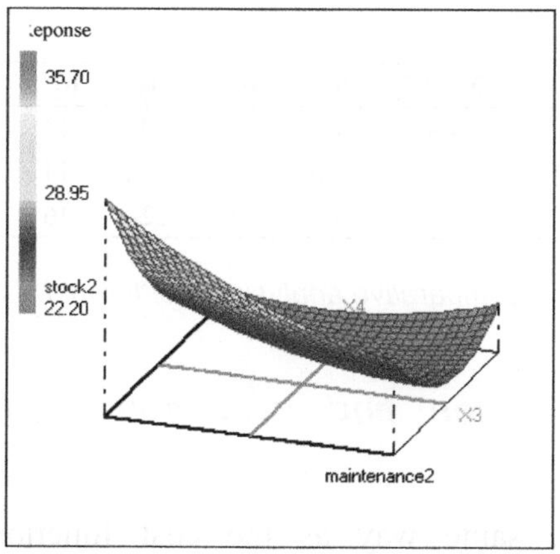

Figure 4.4. *Variation of the cost function in space (h_1,m_2, h_2). For a color version of this figure, see www.iste.co.uk/rezg/services.zip*

In the experimental field, function Π_1 varies between 21.54 and 36.6 mu.

Tables 4.4 and 4.5 show the results of an analysis of the sensitivity of the optimal solution incorporating the variations of some of the parameters in our study. Table 4.4 shows the variation of the solution following the variation in the C_p/C_s (shortage cost/storage cost) ratio. The higher this ratio rises – if there is an increase in shortage cost or a decrease in storage cost – the higher the optimal values of stock levels are. In Table 4.5 we analyze the effect of variation of the M_c/M_p (cost of a corrective maintenance/cost of a preventive maintenance) ratio. The increase in this ratio causes a fall in maintenance periodicities and an increase in total cost.

	C_p/C_s	m_1*	h_1*	m_2*	h_2*	Π_1*(m_i*, h_i*)
	50	197	1	223	42	20.95
Initial reference	65	197	1	223	43	21.54
	100	197	2	224	45	23.02
	200	196	3	225	48	24.97

Table 4.4. *Comparative analysis for different Cp/Cs ratios*

	M_C/M_P	m_1*	h_1*	m_2*	h_2*	Π_1*(m_i*, h_i*)
	6.6	197	1	223	43	21.54
Initial reference	10	196	2	222	44	23.02
	20	195	3	220	46	24.97

Table 4.5. *Comparative analysis for different MC/MP ratios*

4.5.3. *Determination and validation of the availability function*

Obtained in the same way as the cost function, the stationary availability function (Π_2) is associated with the data used as follows:

$$\Pi_2 = 0.710713 + 0.111426*X_{m1} + 0.000716667*X_{h1} + 0.114283*X_{m2} + 0.00206111*X_{h2} - 0.143572*X_{m1}^2 + 0.00044375*X_{m1}*X_{h1} + 0.0270563*X_{m1}*X_{m2} + 0.00205625*X_{m1}*X_{h2} - 0.00285577*X_{h1}^2 - 0.00455625*X_{h1}*X_{m2} - 0.00205625*X_{h1}*X_{h2} - 0.127856*X_{m2}^2 + 0.00205625*X_{m2}*X_{h2} + 0.0221442*X_{h2}^2$$

The analysis of variance for each coefficient calculated by the STATGRAPHICS software package is given in Table 4.6.

Source	SCE	ddl	Variance	% Cont.	$F_{calculated}$	Test
$\alpha_1 : m_1$	**0.201232**	**1**	**0.201232**	**38.30**	**465.4**	**S**
$\alpha_2 : h_1$	0.000009245	1	0.000009245	0.002	0.02	NS
$\alpha_3 : m_2$	**0.235092**	**1**	**0.235092**	**44.73**	**543.75**	**S**
$\alpha_4 : h_2$	0.0000764672	1	0.000076467	0.01	0.18	NS
$\alpha_{11} : m_1^2$	**0.0322772**	**1**	**0.0322772**	**6.13**	**74.65**	**S**
$\alpha_{12} : m_1.h_1$	0.0000031506	1	0.00000315063	0.00006	0.01	NS
$\alpha_{13} : m_1.m_2$	**0.0117127**	**1**	**0.0117127**	**2.22**	**27.09**	**S**
$\alpha_{14} : m_1.h_2$	0.0000676506	1	0.0000676506	0.02	0.16	NS
$\alpha_{22} : h_1^2$	0.0000189887	1	0.0000189887	0.0036	0.04	NS
$\alpha_{23} : h_1.m_2$	0.000332151	1	0.000332151	0.06	0.77	NS
$\alpha_{24} : h_1.h_2$	0.0000676506	1	0.0000676506	0.01	0.16	NS
$\alpha_{33} : m_2^2$	**0.0380619**	**1**	**0.0380619**	**7.24**	**88.03**	**S**
$\alpha_{34} : m_2.h_2$	0.0000676506	1	0.0000676506	0.01	0.16	NS
$\alpha_{44} : h_2^2$	0.00114175	1	0.00114175	0.22	2.64	NS
Residues	0.00518824	12	0.000432354	0.97		
Σ	0.5253	26		100.0		

Table 4.6. *Results of the analysis of variance*

As could be expected, the only significant effects are those related to maintenance periodicities in the case of availability, which means that they are significantly different from zero with a 95% confidence level.

After the "pooling" operation, the equation becomes:

$$\Pi_{20} = 0.710713 + 0.111426*X_{m1} + 0.114283*X_{m2} - 0.143572*X_{m1}^2 + 0.0270563*X_{m1}*X_{m2} - 0.127856*X_{m2}^2$$

The optimal values, X_{m1}* and X_{m2}*, are obtained by solving the following equations:

$$\left.\frac{\partial \Pi_2}{\partial X_{m1}}\right|_{X_{m1}=X^*_{m1}} = 0$$

and

$$\left.\frac{\partial \Pi_2}{\partial X_{m2}}\right|_{X_{m2}=X^*_{m2}} = 0$$

hence $X_{m1}* = 0.44578$ and $X_{m2}* = 0.515639$.

Returning to real values, we obtain $m_1* = 298$ time units and $m_2* = 294$ time units, giving a maximum availability of 79%.

Figures 4.5 and 4.6 show the variation in availability in relation to maintenance periodicities.

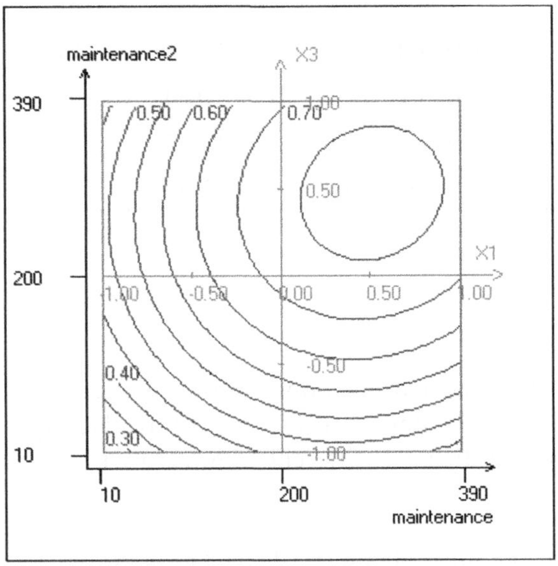

Figure 4.5. *Iso-responses of the availability function in plane (m_1, m_2). For a color version of this figure, see www.iste.co.uk/rezg/services.zip*

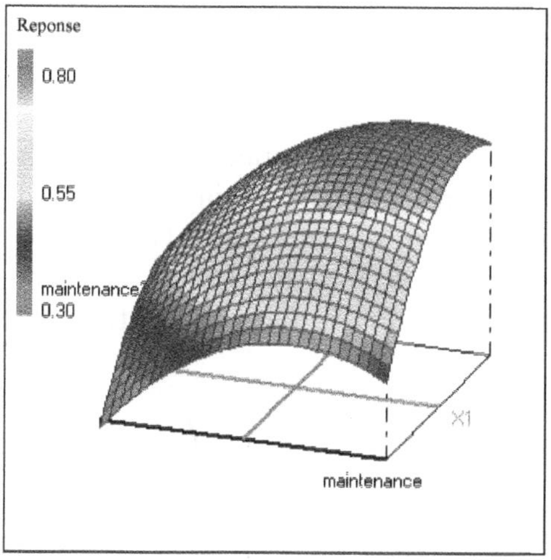

Figure 4.6. *Variation of the availability function in plane (m_1, m_2). For a color version of this figure, see www.iste.co.uk/rezg/services.zip*

In the experimental field, function Π_2 varies between 24% and 79%.

4.6. Finding the best compromise between cost, availability and quality: multi-criteria analysis

Our control policy aims to determine the best compromise taking into account the cost criterion (to be minimized), that of the stationary availability of the system (to be maximized) and that of the number of parts rejected (to be minimized).

The number of parts lost *N*(*t*) follows a model inspired by Rosenblatt's formula [ROS 85]:

$$N(t)\begin{cases} 0 & \text{if } t \leq \tau \\ \alpha.P(t-\tau) & \text{if } t > \tau \end{cases}$$

where:

– α is the percentage of faulty parts during a production cycle;

– τ is the age of beginning of breakdown;

– P is the production rate.

From the formula above, we notice that the number of parts lost is null until instant *t*, representing beginning of a breakdown. We wish to set the objective of a zero rejection rate. This implies that the time to preventive maintenance m_i (i = 1, 2) must be within an upper limit equal to τ. For our example, τ is equal to 120 time units.

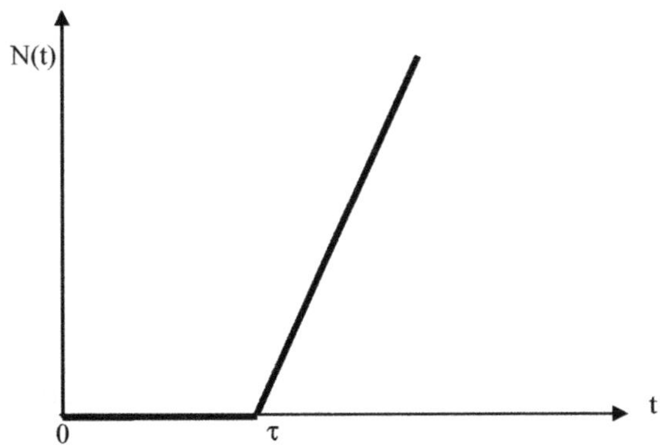

Figure 4.7. *Variation of function N(t)*

It becomes quite clear that the ideal solution permitting us to simultaneously reach optimal values for the three criteria considered – minimum cost, maximum availability and minimal loss – cannot be reached. The levels of the factors for an optimal response are not the same as those that optimize the other responses.

The notion of desirability [HAR 65] enables us to seek a compromise between different objectives that may be weighted. To apply this concept, the Deringer representation [DER 80] is used (see Figure 4.8) in order to graphically display the objectives of each response.

To determine the optimal value for the target considered, the experimenter sets an acceptable upper or lower limit. If the response obtained is equal to the target value, we will say that the partial desire of the experimenter for that objective is 1 (goal 100% achieved). If the response obtained is higher or equal to the upper limit, we will say that the partial desire of the experimenter for that objective is equal to zero (goal 0% achieved). Between these two points, Deringer suggests modeling the evolution through a curve whose equation is given in Figure 4.8. We will take a high T-shaped coefficient when this function is preferred (Figure 4.9). We will take a low T-shaped coefficient when the deviation of the response from the target value is less important (Figures 4.10 and 4.11).

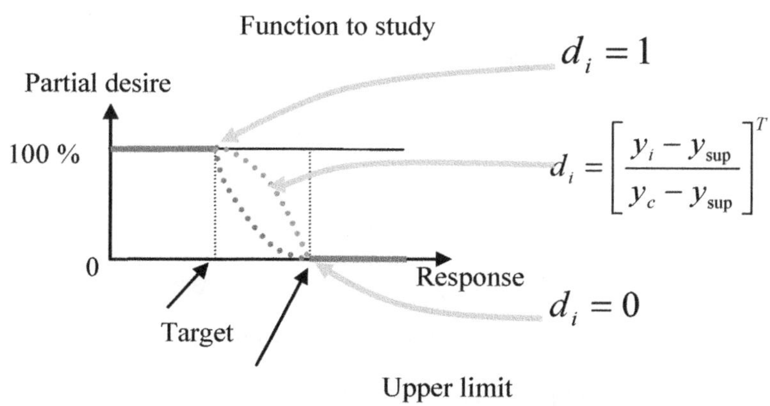

Figure 4.8. *Representation of desirability. For a color version of this figure, see www.iste.co.uk/rezg/services.zip*

In the case of the situation analyzed, where the minimum cost is 21.54 mu, let us suppose, for example, that the decision-maker accepts an upper cost limit of 23 μ if this increase benefits availability.

The decision-maker wants to know whether availability is above 70%, as the target value is a maximum availability that of 79%.

As for the loss, the ideal goal being 0%, the decision-maker sets an acceptable upper limit at 5%.

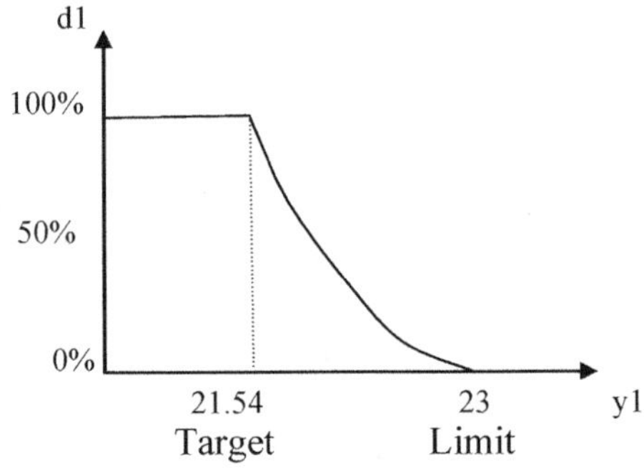

Figure 4.9. *Desirable cost*

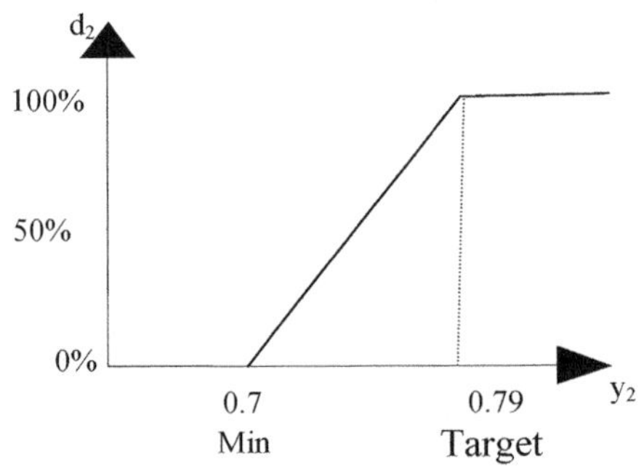

Figure 4.10. *Desirable level of availability*

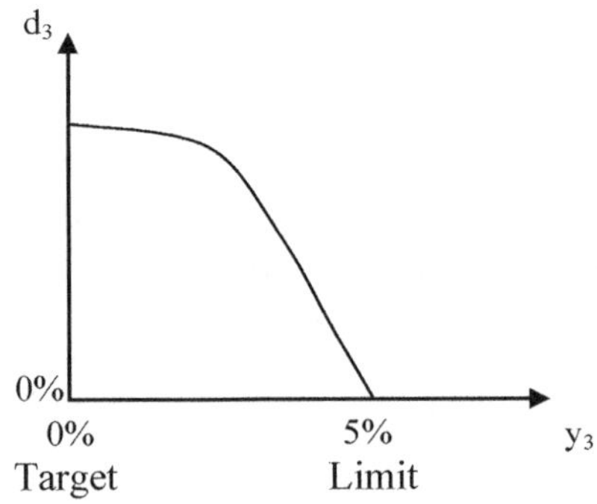

Figure 4.11. *Desirable level of loss*

The shape of the curves for partial desire d_i being defined for each objective, they are turned into a global desire function D, in which each partial desire is weighted by a weight W_i, depending on its relative importance.

This function D is defined by:

$$D = \sqrt[w]{d_1^{w_1} . d_2^{w_2} ... d_m^{w_m}}$$

with:

$$w = \sum_{i=1}^{m} w_i$$

The function obtained this way being too complex to be analytically optimized, only numerical optimization procedures can be used. This process is proposed by the computation software package NEMROD (LPRAI, Marseille). This package has enabled us to achieve a compromise (see Table 4.7) by reaching:

– 87% of the minimum cost objective, with a value of 21.66 mu;

– 76% of the maximum availability objective, with a value of 77%;

– 56% of the minimum loss objective, with a value of 4%.

Table 4.8 summarizes the results obtained.

Response	Response name	Value	d_i %	Coef. T	Weight w_i
y_1	Mean cost	21.66	87.65	10	2
y_2	Availability	0.77	76.13	1	2
y_3	Loss	4%	56.2	0.5	1
	DESIRABILITY		81.69		

Table 4.7. *The compromise reached*

	Minimum cost	Availability	Losses	Compromise determined
m_1 (t.u)	197	298	120	229
h_1 (unit)	1	/	/	1
m_2 (t.u)	223	294	120	240
h_2 (unit)	40	/	/	40

Table 4.8. *Summary of results*

4.7. Conclusion

This chapter has dealt with the determination of the operational variables of a production system made up of two machines in series and two buffer stocks. This system makes only one type of product with a certain rejection rate because of non-conformity. The failure rate of each machine increases over time. Machines undergo age-type maintenance with random lengths between interventions. The decision variables considered are the ages at which each machine must undergo preventive maintenance, and the maximum levels of the two buffer stocks. The huge difficulty of analytically solving this problem has been overcome by using an approach based on simulation and experimental design, accompanied by multi-criteria analysis.

To achieve this we have developed a simulation model based on an integrated maintenance strategy pairing maintenance management and stock management through control. The techniques arising from experimental design permit us to derive a formal model to represent the cost function and the system availability function. These models represent a good approximation of the cost and availability functions. They are models that are easy to implement and optimize.

As both functions of cost and availability obtained were simple quadratic functions, the optimal values of the decision variables were easily determined. Thanks to the implementation of multi-criteria analysis based on the notion of desirability, we have found an area of compromise between three criteria:

– cost (to be minimized);

– availability (to maximized);

– loss (to be minimized);

while simultaneously optimizing these responses.

The path is thus clear for generalising the approach recommended in this chapter to situations involving *m* machines and *n* stocks, which will now be further investigated.

Appendices

Appendix 1

It is assumed that the demand variable has its first and second statistic moments perfectly known for each period k, that is, $E\{d_k\} = \hat{d}_k$ (mean) and $Var_{d_k} = \sigma_d^2$ (standard deviation) for each k.

The inventory variable S_k is statistically described by its mean; $E\{S_k\} = \hat{S}_k$ and its variance Var_{S_k} $Var_{S_k} = E\{(S_k - \hat{S}_k)^2\}$:

$$Var_{S_k} = E\left\{(S_k - \hat{S}_k)^2\right\}$$

U_k being constant for each interval Δt, we have $\hat{U}_k = U_k$ and $Var_{U_k} = 0$.

U_k is essentially deterministic, since it does not depend on the random variables and $\hat{S}_k$.

Thus $\hat{U}_k = U_k$ and $Var_{U_k} = 0$.

Balance equation [1.2] can be reformulated as follows:

$$\hat{S}_{k+1} = \hat{S}_k + U_{1,k} + \beta_2 \cdot U_{2,k-\tau} - \hat{d}_k \qquad \text{[A1.1]}$$

If we determine the difference between equations [1.2] and [A1.1], we obtain:

$$S_{k+1}-\hat{S}_{k+1}=S_k-\hat{S}_k-(d_k-\hat{d}_k)$$

$$\Rightarrow (S_{k+1}-\hat{S}_{k+1})^2=((S_k-\hat{S}_k)-(d_k-\hat{d}_k))^2$$

$$\Rightarrow E\left\{(S_{k+1}-\hat{S}_{k+1})^2\right\}=E\left\{((S_k-\hat{S}_k)-(d_k-\hat{d}_k))^2\right\}$$

$$\Rightarrow E\left\{(S_{k+1}-\hat{S}_{k+1})^2\right\}=E\left\{(S_k-\hat{S}_k)^2\right\}+E\left\{(d_k-\hat{d}_k)^2\right\}-2E\left\{(S_k-\hat{S}_k)(d_k-\hat{d}_k)\right\}$$

Since S_k and d_k are independent random variables, we can deduce that:

$$E\left\{(S_k-\hat{S}_k)(d_k-\hat{d}_k)\right\}=E\left\{(S_k-\hat{S}_k)\right\}E\left\{(d_k-\hat{d}_k)\right\}$$

Hence:

$$E\left\{(S_k-\hat{S}_k)\right\}=E\{S_k\}-E\left\{\hat{S}_k\right\}=0$$

and

$$E\left\{(d_k-\hat{d}_k)\right\}=E\{d_k\}-E\left\{\hat{d}_k\right\}=0$$

Therefore

$$E\left\{(S_{k+1}-\hat{S}_{k+1})^2\right\}=E\left\{(S_k-\hat{S}_k)^2\right\}+E\left\{(d_k-\hat{d}_k)^2\right\}$$

$$\Rightarrow (\sigma_{S_{k+1}})^2=(\sigma_{S_k})^2+(\sigma_{d_k})^2 \qquad \text{[A1.2]}$$

If we assume that $\sigma_{S_0}=0$ (for $k=0$) and that σ_{d_k} is constant and equal to σ_d for all periods, we can deduce that:

$$(\sigma_{S_k})^2=k(\sigma_{d_k})^2$$

Since: $Var_{S_k} = E\left\{(S_k - \hat{S}_k)^2\right\} = E\left\{S_k{}^2\right\} - \hat{S}_k^2$

and: $Var_{S_k} = (\sigma_{S_k})^2 = k(\sigma_{d_k})^2$

we can write:

$$E\left\{S_k{}^2\right\} - \hat{S}_k^2 = k(\sigma_d)^2$$

$$E\left\{S_k{}^2\right\} = k(\sigma_d)^2 + \hat{S}_k^2 \qquad [A1.3]$$

Substituting [A1.3] in expression [1.6] of the expected cost, we obtain:

$$F(u) = C_s \cdot \hat{S}_H^2 + \sum_{k=0}^{H-1}\left[C_s \cdot \hat{S}_k^2 + C_{pr1} \cdot U_{1,k}^2 + C_{pr2} \cdot \beta_2 \cdot U_{2,k}^2\right] + C_s \cdot (\sigma_d)^2 \cdot \sum_{k=0}^{H-1} k$$

$$F(u) = C_s \cdot \hat{S}_H^2 + \sum_{k=0}^{H-1}\left[C_s \cdot \hat{S}_k^2 + C_{pr1} \cdot U_{1,k}^2 + C_{pr2} \cdot \beta_2 \cdot U_{2,k}^2\right] + C_s \cdot (\sigma_d)^2 \cdot \frac{H(H+1)}{2}$$

Appendix 2

PROOF FOR LEMMA A2.1.–

Since the demand is assumed to follow Gaussian distribution, it is assumed that the first and second static moments of the demand variable are known.

For each period k, if the expected demand is equal to the average demand, we then have $E[d(k)] = \hat{d}(k)$ and $V_k = \sigma_k^2$ with $\forall k \in \{0, 1, 2, ..., H\}$.

The inventory variables $S_1(k)$ and $S_0(k)$ are statistically described by their means and variances, thus the expected inventory level at S_1 is equal to the average inventory level at S_1 and the expected inventory level at S_0 is equal to the average inventory level at S_0. We then have:

$$E[S_1(k)] = \hat{S}_1(k), \quad E[S_0(k)] = \hat{S}_0(k)$$

$$E\left[\left(S_1(k) - \hat{S}_1(k)\right)^2\right] = V_{S_1,k}, \quad E\left[\left(S_0(k) - \hat{S}_0(k)\right)^2\right] = V_{S_0,k}$$

where $V_{S_1,k}$ and $V_{S_0,k}$ are the variance in the inventory of S_1 and S_0, respectively.

The control variables, $u(k)$ and $Q(k)$, are deterministic for each interval Δt. Therefore, $V_{u,k} = V_{Q,k} = 0$, with $V_{u,k}$ and are a variation of $u(k)$ and $Q(k)$, respectively.

Where $\hat{r}(k) = \delta \times \hat{d}(k - \tau_r)$, the inventory balance equations [2.2] and [2.3] can be converted as follows:

$$\Rightarrow E[S_1(k+1)] = E[S_1(k) + u(k) - Q(k)] \qquad [2.2]$$

$$\Rightarrow \hat{S}_1(k+1) = \hat{S}_1(k) + \hat{u}(k) - \hat{Q}(k)$$

with

$$k = \{0, 1, \ldots, H\text{-}1\} \qquad [A2.1]$$

$$\Rightarrow E[S_0(k+1)] = \begin{cases} E[S_0(k) + Q(k-\tau) - d(k) + r(k)] & \text{If } k \geq \tau \\ E[S_0(k) - d(k) + r(k)] & \text{otherwise} \end{cases} \qquad [2.3]$$

$$\Rightarrow \hat{S}_0(k+1) = \begin{cases} \hat{S}_0(k) + \hat{Q}(k-\tau) - \hat{d}(k) + \hat{r}(k) & \text{If } k \geq \tau \\ \hat{S}_0(k) - \hat{d}(k) + \hat{r}(k) & \text{otherwise} \end{cases}$$

with

$$k = \{0, 1, \ldots, H\text{-}1\} \text{ and } \tau \geq 1. \qquad [A2.2]$$

Equations [A2.1] and [A2.2] represent the mean inventory in each period k, with $k = \{0, 1, \ldots, H-1\}$.

Taking the difference between [2.2] and [A2.1], we have:

$$S_1(k+1) - \hat{S}_1(k+1) = S_1(k) - \hat{S}_1(k)$$

$$\Rightarrow \left(S_1(k+1) - \hat{S}_1(k+1)\right)^2 = \left(S_1 - \hat{S}_1(k)\right)^2$$

$$\Rightarrow E\left[\left(S_1(k+1) - \hat{S}_1(k+1)\right)^2\right] = E\left[\left(S_1 - \hat{S}_1(k)\right)^2\right]$$

Consequently,

$$V_{S_1,(k+1)} = V_{S_1,k} \qquad k \geq 0$$

Taking the difference between [2.3] and [A2.2], we have:

$$E\left[S_0(k+1)\right] - \hat{S}_0(k+1)$$

Based on equations [2.3], [2.4] and [A2.2], and according to the values of τ and τ_r, we have four cases:

– case 1 when $k \geq \tau_r$ and $k \geq \tau$;

– case 2 when $k \geq \tau_r$ and $k<\tau$;

– case 3 $k < \tau_r$ and $k \geq \tau$;

– case 4 when $k < \tau_r$ and $k<\tau$.

1) Case 1:

When $k \geq \tau_r$ and $k \geq \tau$, we have:

$$S_0(k+1) = S_0(k) + Q(k-\tau) + \delta \times d(k-\tau_r) - d(k)$$

and

$$\hat{S}_0(k+1) = \hat{S}_0(k) + Q(k-\tau) + \delta \times \hat{d}(k-\tau_r) - \hat{d}(k)$$

The difference between [2.3] and [A2.2] is given by:

$$S_0(k+1) - \hat{S}_0(k+1) = (S_0(k) + Q(k-\tau) - d(k) + \delta \times d(k-\tau_r))$$

$$-\left(\hat{S}_0(k) + \hat{Q}(k-\tau) - \hat{d}(k) + \delta \times \hat{d}(k-\tau_r)\right)$$

$$= \left(S_0(k) - \hat{S}_0(k)\right) - \left(d(k) + \hat{d}(k)\right) + \delta \times \left(d(k-\tau_r) - \hat{d}(k-\tau_r)\right)$$

Then the difference between [2.3] and [A2.2] in quadratic form is given by:

$$\left(S_0(k+1)-\hat{S}_0(k+1)\right)^2=\left(S_0(k)-\hat{S}_0(k)\right)^2+\left(d(k)+\hat{d}(k)\right)^2+\delta^2\times\left(d(k-\tau_r)-\hat{d}(k-\tau_r)\right)^2$$
$$-2\times\left(S_0(k)-\hat{S}_0(k)\right)\times\left(d(k)+\hat{d}(k)\right)+2\times\delta\times\left(S_0(k)-\hat{S}_0(k)\right)\times\left(d(k-\tau_r)-\hat{d}(k-\tau_r)\right)-$$
$$2\times\delta\times\left(d(k)+\hat{d}(k)\right)\times\left(d(k-\tau_r)-\hat{d}(k-\tau_r)\right)$$

Then we have the expected difference between [2.3] and [A2.2] in quadratic form that is given by:

$$E\left[\left(S_0(k+1)-\hat{S}_0(k+1)\right)^2\right]=E\left[\left(S_0(k)-\hat{S}_0(k)\right)^2\right]+E\left[\left(d(k)+\hat{d}(k)\right)^2\right]+\delta^2\times E\left[\left(d(k-\tau_r)-\hat{d}(k-\tau_r)\right)^2\right]$$
$$-2\times E\left\{\left(S_0(k)-\hat{S}_0(k)\right)\times\left(d(k)+\hat{d}(k)\right)\right\}+2\times\delta\times E\left[\left(S_0(k)-\hat{S}_0(k)\right)\times\left(d(k-\tau_r)-\hat{d}(k-\tau_r)\right)\right]-$$
$$2\times\delta\times E\left[\left(d(k)+\hat{d}(k)\right)\times\left(d(k-\tau_r)-\hat{d}(k-\tau_r)\right)\right]$$

Since $S_0(k)$ is an independent random variable, we can deduce that:

$$E\left[S_0(k)\right]=E\left[\hat{S}_0(k)\right]$$

So we have:

$$E\left[\left(S_0(k)-\hat{S}_0(k)\right)\times\left(d(k)+\hat{d}(k)\right)\right]=E\left[\left(S_0(k)-\hat{S}_0(k)\right)\right]\times E\left[\left(d(k)+\hat{d}(k)\right)\right]$$

$$E\left[\left(S_0(k)-\hat{S}_0(k)\right)\times\left(d(k-\tau_r)-\hat{d}(k-\tau_r)\right)\right]=E\left[\left(S_0(k)-\hat{S}_0(k)\right)\right]\times E\left[\left(d(k-\tau_r)-\hat{d}(k-\tau_r)\right)\right]$$
$$E\left[\left(d(k)+\hat{d}(k)\right)\times\left(d(k-\tau_r)-\hat{d}(k-\tau_r)\right)\right]=E\left[\left(d(k)+\hat{d}(k)\right)\right]\times E\left[\left(d(k-\tau_r)-\hat{d}(k-\tau_r)\right)\right]$$

hence:

$$E\left[\left(S_0(k)-\hat{S}_0(k)\right)\right]=E[S_0(k)]\times E\left[\hat{S}_0(k)\right]=0$$

$$E\left[d(k-\tau_r)-\hat{d}(k-\tau_r)\right]=E[d(k-\tau_r)]-E\left[\hat{d}(k-\tau_r)\right]=\hat{d}(k-\tau_r)-\hat{d}(k-\tau_r)=0$$

therefore:

$$E\left[\left(S_0(k+1)-\hat{S}_0(k+1)\right)^2\right]=E\left[\left(S_0(k)-\hat{S}_0(k)\right)^2\right]+E\left[\left(d(k)+\hat{d}(k)\right)^2\right]+\delta^2\times E\left[\left(d(k-\tau_r)-\hat{d}(k-\tau_r)\right)^2\right]$$

Then we have the variance of $S_0(k)$, which is given by:

$$V_{S_0,(k+1)} = V_{S_0,k} + V_k + \delta^2 \cdot V_{k-\tau_r} \qquad k \geq \tau_r \ , \ \tau_r \geq 1$$

with

$$V_{S_0,k} = \left(k + (k-1) \times \delta^2\right) \times \sigma_k^2 \qquad k \geq \tau_r \ , \ \tau_r \geq 1$$

The general expression of $V_{S_0,k}$ is determined as follows. Starting with $V_{S_0,(k=0)} = 0$, the variance at time 0 is equal to zero. For $k = 0$:

$$V_{S_0,(k+1)} = V_{S_0,k} + V_k = V_{S_0,1} = V_{S_0,0} + \sigma_k^2 = V_{S_0,0}$$

REMARK.– We know in case 1 that $k \geq \tau \geq 1$. We assumed that $k = 0$ just to determine the general expression $V_{S_0,k}$.

For $k = 1$:

$$V_{S_0,2} = V_{S_0,1} + V_1 + \delta^2 \times V_{1-\tau_r} \qquad \tau_r \geq 1$$

$$V_{S_0,2} = \left(2 + \delta^2\right) \times \sigma_k^2$$

For $k \geq 2$:

$$V_{S_0,(k+1)} = V_{S_0,k} + V_k + \delta^2 \times V_{k-\tau_r} \qquad \tau_r \geq 1$$

$$V_{S_0,k} = \left(k + (k-1) \times \delta^2\right) \times \sigma_k^2 \qquad \tau_r \geq 1 \ , \ k \geq 2$$

The general expression of $V_{S_0,k}$ is given by:

$$V_{S_0,k} = \begin{cases} \left(k + (k-1)\times\delta^2\right)\times\sigma_k^2 & \text{If } k \geq 1 \\ 0 & \text{If } k = 0 \end{cases}$$

Since:

$$\Rightarrow E\left[\left(S_0(k)-\hat{S}_0(k)\right)^2\right] = E\left[S_0(k)^2\right] - \hat{S}_0(k)^2$$

$$\Rightarrow E\left[S_0(k)^2\right] - \hat{S}_0(k)^2 = \left(k+(k-1)\times\delta^2\right)\times\sigma_k^2 \quad k \geq \tau_r \text{ , } \tau_r \geq 1$$

Finally we have:

$$E\left[S_0(k)^2\right] = \left(k+(k-1)\times\delta^2\right)\times\sigma_k^2 + \hat{S}_0(k)^2 \qquad \text{[A2.3]}$$

if $k \geq \tau_r$ and $k \geq \tau$.

2) Case 2:

If $k \geq \tau_r$ and $k < \tau$, we have:

$$S_0(k+1) = S_0(k) + \delta\times d(k-\tau_r) - d(k)$$

and:

$$\hat{S}_0(k+1) = \hat{S}_0(k) + \delta\times\hat{d}(k-\tau_r) - \hat{d}(k)$$

then the difference between [2.3] and [A2.2] is given by:

$$S_0(k+1) - \hat{S}_0(k+1) = \left(S_0(k) + Q(k-\tau) - d(k) + \delta\times d(k-\tau_r)\right)$$

$$-\left(\hat{S}_0(k) + \hat{Q}(k-\tau) - \hat{d}(k) + \delta\times\hat{d}(k-\tau_r)\right)$$

$$= \left(S_0(k) - \hat{S}_0(k)\right) - \left(d(k) + \hat{d}(k)\right) + \delta\times\left(d(k-\tau_r) - \hat{d}(k-\tau_r)\right)$$

If this case is same as case 1, then we have:

$$E\left[S_0(k)^2\right] = \left(k + (k-1)\times\delta^2\right)\times\sigma_k^2 + \hat{S}_0(k)^2 \qquad [A2.4]$$

if $k \geq \tau_r$ and $k < \tau$.

3) Case 3:

If $k < \tau_r$ and $k \geq \tau$, we have:

and:

$$\hat{S}_0(k+1) = \hat{S}_0(k) + Q(k-\tau) - \hat{d}(k)$$

Then the difference between [2.3] and [A2.2] is given by:

$$S_0(k+1) - \hat{S}_0(k+1) = \left(S_0(k) - \hat{S}_0(k)\right) - \left(d(k) + \hat{d}(k)\right)$$

Then we have the difference between [2.3] and [A2.2] in quadratic form, which is given by:

$$\left(S_0(k+1) - \hat{S}_0(k+1)\right)^2 = \left(S_0(k) - \hat{S}_0(k)\right)^2 + \left(d(k) + \hat{d}(k)\right)^2$$
$$-2\cdot\left(S_0(k) - \hat{S}_0(k)\right)\times\left(d(k) + \hat{d}(k)\right)$$

Then we have the expected difference between [2.3] and [A2.2] in quadratic form, given by:

$$E\left[\left(S_0(k+1) - \hat{S}_0(k+1)\right)^2\right] = E\left[\left(S_0(k) - \hat{S}_0(k)\right)^2\right] + E\left[\left(d(k) + \hat{d}(k)\right)^2\right]$$
$$-2\times E\left[\left(S_0(k) - \hat{S}_0(k)\right)\times\left(d(k) + \hat{d}(k)\right)\right]$$

Hence:

$$E\left[d(k) + \hat{d}(k)\right] = E[d(k)] - E\left[\hat{d}(k)\right] = 0$$

$$V_{S_0,(k+1)} = V_{S_0,k} + V_k$$

with:

$$V_{S_0,k} = \begin{cases} \left(k + (k-1)\times\delta^2\right)\times\sigma_k^2 & \textit{If } k \geq 1 \\ 0 & \textit{If } k = 0 \end{cases}$$

In this case, if we have $k < \tau_r$ and $k \geq \tau \geq 1$, then we have:

$$V_{S_0,k} = \left(k + (k-1)\times\delta^2\right)\times\sigma_d^2$$

Since:

$$\Rightarrow E\left[\left(S_0(k) - \hat{S}_0(k)\right)^2\right] = E\left[S_0(k)^2\right] - \hat{S}_0(k)^2$$

$$\Rightarrow E\left[S_0(k)^2\right] - \hat{S}_0(k)^2 = \left(k + (k-1)\times\delta^2\right)\times\sigma_k^2$$

Thus:

$$E\left[S_0(k)^2\right] = \left(k + (k-1)\times\delta^2\right)\times\sigma_k^2 + \hat{S}_0(k)^2 \qquad \text{[A2.5]}$$

if $k < \tau_r$ and $k < \tau$.

4) Case 4:

If $k<\tau_r$ and $k<\tau$, we have:

$$S_0(k+1) = S_0(k) - d(k)$$

and:

$$\hat{S}_0(k+1) = \hat{S}_0(k) - \hat{d}(k)$$

Then the difference between [2.3] and [A2.2] is given by:

$$S_0(k+1) - \hat{S}_0(k+1) = \left(S_0(k) - \hat{S}_0(k)\right) - \left(d(k) + \hat{d}(k)\right)$$

Then the difference between [2.3] and [A2.2] in quadratic form is given by:

$$\left(S_0(k+1)-\hat{S}_0(k+1)\right)^2=\left(S_0(k)-\hat{S}_0(k)\right)^2+\left(d(k)+\hat{d}(k)\right)^2$$
$$-2\cdot\left(S_0(k)-\hat{S}_0(k)\right)\times\left(d(k)+\hat{d}(k)\right)$$

Then we have the expected value of the difference between [2.3] and [A2.2] in quadratic form, which is given by:

$$E\left[\left(S_0(k+1)-\hat{S}_0(k+1)\right)^2\right]=E\left[\left(S_0(k)-\hat{S}_0(k)\right)^2\right]+E\left[\left(d(k)+\hat{d}(k)\right)^2\right]$$
$$-2\times E\left[\left(S_0(k)-\hat{S}_0(k)\right)\times\left(d(k)+\hat{d}(k)\right)\right]$$

Hence:

$$E\left[\left(d(k)+\hat{d}(k)\right)\right]=E[d(k)]-E\left[\hat{d}(k)\right]=0$$

$$V_{S_0,(k+1)}=V_{S_0,k}+V_k$$

with:

$$V_{S_0,k}=\begin{cases}\left(k+(k-1)\times\delta^2\right)\times\sigma_k^2 & \text{If } k\geq 1\\ 0 & \text{If } k=0\end{cases}$$

Since:

$$\Rightarrow E\left[\left(S_0(k)-\hat{S}_0(k)\right)^2\right]=E\left[S_0(k)^2\right]-\hat{S}_0(k)^2$$

$$\Rightarrow E\left[S_0(k)^2\right]-\hat{S}_0(k)^2=\begin{cases}\left(k+(k-1)\times\delta^2\right)\times\sigma_k^2 & k\geq 1\\ 0 & k=0\end{cases}$$

thus if $k<\tau_r$ and $k<\tau$ we have:

$$E\left[S_0(k)^2\right]=\begin{cases}\left(k+(k-1)\times\delta^2\right)\times\sigma_k^2+\hat{S}_0(k)^2 & k\geq 1\\ \hat{S}_0(0)^2 & k=0\end{cases}\qquad\text{[A2.6]}$$

In conclusion in these four cases, for all $k = \{0, 1,\ldots, H-1\}$, $\tau_r \geq 1$ and $\tau \geq 1$, we have:

$$E\left[S_0(k)^2\right] = \begin{cases} \left(k + (k-1)\times\delta^2\right)\times\sigma_k^2 + \hat{S}_0(k)^2 & k \geq 1 \\ \hat{S}_0(0)^2 & k = 0 \end{cases} \quad \text{[A2.7]}$$

Substituting [A2.7] in the expected cost [2.11], we obtain:

$$GT_{USQ} = \sum_{k=0}^{H}\left(C_{S1}\times\hat{S}_1(k)^2 + C_{S0}\times\hat{S}_0(k)^2\right) + \sum_{k=1}^{H}\left(\left(k+(k-1)\times\delta^2\right)\times\sigma_k^2\right) + \sum_{k=0}^{H-1}\left(C_{pr}\times u(k)^2 + C_l\times\left(\frac{Q(k)}{Q_v}\right)^2\right)$$

$$= \sum_{k=0}^{H}\left(C_{S1}\times\hat{S}_1(k)^2 + C_{S0}\times\hat{S}_0(k)^2\right) + C_{S0}\times\sum_{k=1}^{H}\left(\sigma_k^2\right)\times\left[H+1+(H-1)\times\delta^2\right]\times\frac{H}{2} + \sum_{k=0}^{H-1}\left(C_{pr}\times u(k)^2 + C_l\times\left(\frac{Q(k)}{Q_v}\right)^2\right)$$

Appendix 3

PROOF OF EQUATION [3.5].–

The inventory variable, S_k, is statistically described by its mean and variance:

$$E\left\{\left(S_k - \hat{S}_k\right)^2\right\} = Var\left(S_k\right)$$

The expected inventory cost is:

$$C_s \cdot E\left\{S_k^{\,2}\right\} = C_s \cdot \hat{S}_k^{\,2}$$

The balance equation [3.2]:

$$S_{k+1} = S_k + U_k - d_k \; k \in \{0,1,\ldots.,L-1\}$$

can be converted into an equivalent inventory balance equation, as follows

$$\Rightarrow E\left\{S_{k+1}\right\} = E\left\{S_k\right\} + U_k - d_k \qquad [3.2]$$

$$\Rightarrow \hat{S}_{k+1} = \hat{S}_k + U_k - \hat{d}_k \qquad [A3.1]$$

Equation [A3.1] represents the mean variation of inventory at each period k:

$k \in \{1,2,\ldots\ldots,N-1\}$.

Furthermore, $u_{i,k}$ is deterministic, since it does not depend on the random variables d_k and S_k. That is, $E\{U\}=U_k$ *with* $V(U_k)=0$ $\forall k$. Taking the difference between [3.2] and [A.1]:

$$S_{k+1}-\hat{S}_{k+1}=S_k-\hat{S}_k-\left(d_k-\hat{d}_k\right)$$

$$\Rightarrow\left(S_{k+1}-\hat{S}_{k+1}\right)^2=\left(\left(S_k-\hat{S}_k\right)-\left(d_k-\hat{d}_k\right)\right)^2$$

$$\Rightarrow E\left(\left(S_{k+1}-\hat{S}_{k+1}\right)^2\right)=E\left(\left(\left(S_k-\hat{S}_k\right)-\left(d_k-\hat{d}_k\right)^2\right)\right)$$

$$\Rightarrow E\left(\left(S_{k+1}-\hat{S}_{k+1}\right)^2\right)=E\left(\left(S_k-\hat{S}_k\right)^2+\left(d_k-\hat{d}_k\right)^2-2\cdot\left(S_k-\hat{S}_k\right)\cdot\left(d_k-\hat{d}_k\right)\right)$$

$$\Rightarrow E\left(\left(S_{k+1}-\hat{S}_{k+1}\right)^2\right)=E\left(\left(S_k-\hat{S}_k\right)^2\right)+E\left(\left(d_k-\hat{d}_k\right)^2\right)-2\cdot E\left(\left(S_k-\hat{S}_k\right)\cdot\left(d_k-\hat{d}_k\right)\right)$$

Since S_k and d_k are independent random variables, we can deduce that:

$$E\left(\left(S_k-\hat{S}_k\right)\cdot\left(d_k-\hat{d}_k\right)\right)=E\left(\left(S_k-\hat{S}_k\right)\right)\cdot E\left(\left(d_k-\hat{d}_k\right)\right)$$

Also, we can see that:

$$E\left(\left(S_k-\hat{S}_k\right)\right)=E\left(S_k\right)-E\left(\hat{S}_k\right)=0$$

$$E\left(\left(d_k-\hat{d}_k\right)\right)=E\left(d_k\right)-E\left(\hat{d}_k\right)=0$$

Consequently,

$$E\left(\left(S_{k+1}-\hat{S}_{k+1}\right)^2\right)=E\left(\left(S_k-\hat{S}_k\right)^2\right)+E\left(\left(d_k-\hat{d}_k\right)^2\right)$$

If we assume that $\sigma_s(0)=0$ and σ_{dk} is constant and equal to σ_d for all ks, we can deduce that:

$$\left(\sigma_{s_k}\right)^2 = k \cdot (\sigma_d)^2$$
$$\Rightarrow E\left(S_k^2\right) - \hat{S}_k^2 = k \cdot (\sigma_d)^2$$

$$\Rightarrow E\left(S_k^2\right) = k \cdot (\sigma_d)^2 + \hat{S}_k^2 \qquad [A3.2]$$

Substituting [A3.1] in the expected cost [3.1]:

$$Z = C_s \cdot E\left(S_L^2\right) + \sum_{i=0}^{L} C_s \cdot E\left(S_k^2\right) + C_{pr} \cdot U_k^2$$

$$Z(u) = C_s \times \left(\hat{S}_L^2\right) + \sum_{k=0}^{L-1} C_s \cdot \hat{S}_k^2 + C_{pr} \times u_k^2 + C_s \times (\sigma_d)^2 \times \sum_{i=0}^{L} k$$

$$Z(u) = C_s \times \left(\hat{S}_L^2\right) + \sum_{k=0}^{L-1} C_s \cdot \hat{S}_k^2 + C_{pr} \times u_k^2 + C_s \times \sigma_d^2 \times \frac{L(L+1)}{2}$$

PROOF OF EQUATION [3.6].–

$$S(k+1) = S(k) + U(k) - d(k)$$

$$\text{Prob}\left(S(k+1) \geq 0\right) \geq \alpha$$

$$\text{Prob}\left(S(k) + U(k) - d(k) \geq 0\right) \geq \alpha$$

$$\text{Prob}\left(S(k) + U(k) \geq d(k)\right) \geq \alpha$$

$$\text{Prob}\left(S(k) + U(k) - \hat{d}(k) \geq d(k) - \hat{d}(k)\right) \geq \alpha$$

$$\text{Prob}\left(\frac{S(k) + U(k) - \hat{d}(k)}{V_{d,k}} \geq \frac{d(k) - \hat{d}(k)}{V_{d,k}}\right) \geq \alpha$$

where:

– $\hat{d}(k)$ is the average demand at period *k*;

– $Var(d(k)) = V_{d,k} \geq 0$ is the variation in demand *d* during period *k*.

This equation is in the form of Prob [$Y \geq X$]*a*, with $X = \frac{d_k - \hat{d}_k}{V_{d_k}}$ being a Gaussian random variable representing the demand d_k, and φ_{d_k} being a cumulative Gaussian distribution function of the form:

$$\varphi_{d,k}\left(\frac{S(k) + U(k) - \hat{d}(k)}{V_{d,k}}\right) \geq \alpha$$

Since $\lim_{d_k \to -\infty} \varphi_{d_k} = 0$ and $\lim_{d_k \to +\infty} \varphi_{d_k} = 1$, function φ_{d_k} is strictly increasing, and we note that it is indefinitely differentiable. That is why we conclude that φ_{d_k} is invertible.

Thus,

$$\frac{S(k) + U(k) - \hat{d}(k)}{V_{d,k}} \geq \varphi_{d,k}^{-1}(\alpha)$$

$$S(k) + U(k) - \hat{d}(k) \geq V_{d,k} \cdot \varphi_{d,k}^{-1}(\alpha)$$

$$U(k) \geq V_{d,k} \cdot \varphi_{d,k}^{-1}(\alpha) + \hat{d}(k) - S(k) \cdot$$

Bibliography

[AGH 08] AGHZZAF E., NAJID N., "Integrated production and preventive maintenance in production systems subject to random failures", *European Journal of Operational Research*, vol. 178, pp. 3382–3392, 2008.

[AGH 07] AGHEZZAF E.H., JAMALI M.A., AIT-KADI D., "An integrated production and preventive maintenance planning model", *European Journal of Operational Research*, vol. 181, pp. 676–685, 2007.

[AIT 91] AIT-KADI D., CLEROUX C., "Replacement strategies with mixed corrective actions at failure", *Computers and Operations Research Logistic*, vol. 18, no. 2, pp. 141–149, 1991.

[AIT 93] AIT-KADI D., "Availability optimization of randomly failing equipment", in COTSAFTIS M., VERNADAT F. (eds), *Advances in the Factories of the Future, CIM and Robotics*, Elsevier, 1993.

[AKE 86] AKELLA R., KURNAR P.R., "Optimal control of production rate in a failure prone manufacturing system", *IEEE Trans. Automatic Control*, vol. 31, pp. 116–126, 1986.

[AME 01] AMESSE F., DRAGOSTE L., NOLLET J. *et al.*, "Issues on partnering: evidences from subcontracting in aeronautics", *Technovation*, vol. 21, no. 9, pp. 559–569, 2001.

[AND 99] ANDERSEN P.H., "Organizing international technological collaboration in subcontractor relationships: an investigation of the knowledge-stickiness problem", *Research Policy*, vol. 28, no. 6, pp. 625–642, 1999.

[AYE 12] AYED S., DELLAGI S., REZG N., "Joint optimisation of maintenance and production policies considering random demand and variable production rate", *International Journal of Production Research*, vol. 50, no. 23, pp. 6870–6885, 2012.

[BAD 02] BADIA F., BERRADE M.D., CAMPOS, C.A., "Optimal inspection and preventive maintenance of units with revealed and unrevealed failures", *Reliability Engineering and System Safety*, vol. 63, pp. 127–131, 2002.

[BAI 95] BAI S.X., GERSHWIN S.B., "Scheduling manufacturing systems with work-in-process inventory control: Single-part- type systems", *IIE Transactions*, vol. 27, pp. 599–617, 1995.

[BAR 65] BARLOW R.E., PORSCHAN F., *Mathematical Theory of Reliability*, John Wiley & Sons, New York, 1965.

[BER 01] BERTRAND J.W.M., SRIDHARAN V., "A study of simple rules for subcontracting in make-to-order manufacturing", *European Journal of Operational Research*, vol. 128, no 3, pp. 509–531, 2001.

[BER 91] BERKE T.M., ZAINO N.A., "Warranties: what are they? What do they really cost?", *Proceedings of the 1991 IEEE Annual Reliability and Maintainability Symposium*, pp. 326–330, 1991.

[BER 95] BERTESEKAS D.P, "Dynamic programming and optimal control", *Athena Scientific*, vol. II, Belmont, USA, 1995.

[BER 10] BERTHAUTA F., GHARBIA A., KENNEB J.-P. *et al.*, "Improved joint preventive maintenance and hedging point policy", *International Journal of Production Economics*, vol. 127, no. 1, pp. 60–72, 2010.

[BOU 12] BOUGUERRA, S., CHELBI A., REZG N., "A decision model for adopting an extended warranty under different maintenance policies", *International Journal of Production Economics*, vol. 135, pp. 840–849, 2012.

[BUZ 67] BUZACOTT J.A, "Automatic transfer lines with buffer stocks", *International Journal of Production Research*, vol. 5, no. 3, p. 183, 1967.

[BUZ 92] BUZACOTT J.A., SHANTHIKUMAR J.G., *Stochastic models of manufacturing systems*, Prentice-Hall, 1992.

[BUR 95] BURMAN, M.H., New results in flow line analysis, PhD Thesis, MIT, 1995.

[CHE 97] CHEUNG K.L., HAUSMANN W.H., "Joint optimization of preventive maintenance and safety stock in an unreliable production environment", *Naval Research Logistics*, vol. 44, pp. 257–272, 1997.

[CHE 99] CHELBI A., AIT-KADI D., "An optimal Inspection Strategy for randomly failing equipment", *Reliability Engineering and System Safety*, vol. 63, pp 127–131, 1999.

[CHA 01] CHAN F.T., "Simulation analysis of maintenance policies in a flow line production system", *International Journal of Computer Applications in Technology,*, vol.14, nos. 1–3, pp.78–86, 2001.

[CHE 04] CHELBI A., AIT-KADI D., "Analysis of a production/inventory system with randomly failing production unit submitted to regular preventive maintenance", *European Journal of Operational Research,* vol. 156, no. 3, pp. 712–718, 2004.

[CHI 08] CHIEN Y.H., "Optimal age-replacement policy under an imperfect renewing free replacement warranty", *IEEE Transactions on Relability*, vol. 57, no. 1, pp. 125–132, 2008.

[CHI 10] CHIEN Y.H., "Optimal age for preventive replacement under a combined fully renewable free replacement with a pro-rata warranty", *International Journal of Production Economics*, vol. 124, pp. 198–205, 2010.

[CHO 91] CHO D.I., PARLAR M., "A survey of maintenance models for multi-unit systems", *European Journal of Operational Research*, vol. 51, pp. 1–23, 1991.

[COX 62] COX D.R., *Renewal Theory*, Methuen, 1962.

[DAL 92] DALLERY Y., GERSHWIN S.B., "Manufacturing flow line systems: a review of models and analytical results", *Queueing Systems*, vol. 12, pp. 3–94, 1992.

[DAL 99] DALLERY Y., LE BIHAN H., "An improved decomposition method for the analysis of production lines with unreliable machines and finite buffers", *International Journal of Production Research*, vol. 37, no. 5, pp. 1093–1117, 1999.

[DAL 10] DAHANE M., CLEMENTZ C., REZG N., "Effects of extension of subcontracting on a production system in a joint maintenance and production context", *Computers & Industrial Engineering*, vol. 58, no. 1, pp. 88–96, 2010.

[DAS 99] DAS T.K., SARKAR S., "Optimal preventive maintenance in a production inventory system", *IIE Transactions*, vol. 31, pp. 537–551, 1999.

[DEL 07] DELLAGI S., REZG N., UIE U., "Preventive maintenance of manufacturing systems under environmental constraints", *International Journal of Production Research*, vol. 45, no. 5, p. 1233, 2007.

[DER 80] DERRINGER G., SUICH R., "Simultaneous optimization of several response variables", *Journal of Quality Technology*, vol. 12, no. 4, pp. 214–219, 1980.

[DOH 98] DOHI T., YUN W.Y., KAIO N. *et al.*, "Optimal design of economic manufacturing process with machine rate depending on production speed", *Proceedings of the International Symposium on Manufacturing Strategy'98*, pp. 404–409, 1998.

[DOL 02] DOLGUI A., OULD-LOULY M.A., "A model for supply planning under lead timeuncertainty", *International Journal on Production Economics*, vol. 78, pp. 145–152, 2002.

[FAZ 08] FAZEKAS C., SZEDERKENYI G., HANGOS M.K., "Parameter estimation of a simple primary circuit model of a VVER plant", *IEEE Transactions on Nuclear Science*, vol. 55, no. 5, pp. 10, 2008.

[FUN 05] FUNG R.Y.K., CHEN T., "Multiagent supply chain planning and coordination architecture", *International Journal of Advanced Manufacturing Technology*, vol. 25, pp. 811–819, 2005.

[GER 94] GERSHWIN S.B., *Manufacturing Systems Engineering*, PTR Prentice Hall, 1994.

[GHA 00] GHARBI A., KENNE J.P., "Production and corrective maintenance planning of FMS through simulation optimisation", *4th International Conference on Engineering Design and Automation*, Orlando, Florida, USA, 2000.

[GHA 03] GHARBI A., KENNÉ J.P., "Optimal production control problem in stochastic multiple-product multi-machines manufacturing systems", *IIE Transactions*, vol. 35, pp. 941–952, 2003.

[GLA 93] GLASSEY C.R., HONG Y., "Analysis of behaviour of unreliable n-stage transfer line with (n-1) inter-stage storage buffers", *International Journal of Production Research,* vol. 31, no. 3, pp. 519–530, 1993.

[HAJ 09] HAJEJ Z., DELLAGI S., REZG N., "An optimal production/maintenance planning under stochastic random demand, service level and failure rate", *IEEE Explore*, nos. 22-25, pp. 292–297, 2009.

[HAJ 09] HAJEJ, Z., DELLAGI S., REZG N., "An optimal production/maintenance planning under stochastic random demand, service level and failure rate", *IEEE International Conference on Automation Science and Engineering*, Bangalore, pp. 292 –297, 2009.

[HAJ 11] HAJEJ, Z., DELLAGI S., REZG N., "Optimal integrated maintenance/production policy for randomly failing systems with variable failure rate", *International Journal of Production Research*, vol. 49, no. 19, pp. 5695–5712, 2011.

[HAJ 12] HAJEJ Z., DELLAGI S., REZG N., "Joint optimisation of maintenance and production policies with subcontracting and product returns", *Journal of Intelligent Manufacturing*, 2012.

[HAJ 13] HAJEJ Z., REZG N., GHARBI A., "Integrated maintenance policy optimization under lease/warranty contract", *ICIE: International Conference on Industrial Engineering*, Dubai, October 21-23, 2013.

[HAJ 14a] HAJEJ, Z., REZG N., GHARBI A., "Forecasting and maintenance problem under subcontracting constraint with transportation delay", *International Journal of Production Research,* 2014.

[HAJ 14b] HAJEJ Z., DELLAGI S., REZG N., "Joint optimisation of maintenance and production policies with subcontracting and product returns", *Journal of Intelligent Manufacturing*, vol. 25, no. 3, pp. 589–602, 2014.

[HAR 65] HARRINGTON E.C., "The desirability function", *Industrial Quality Control*, vol. 21, no. 10, pp. 494-498, 1965.

[HAX 84] HAX A., CANDEA D., *Production and Inventory Management*, Prentice Hall, 1984.

[HOL 60] HOLT C., MODIGLIANI C., MUTH J.F. *et al.*, *Planning Production, Inventory and Work Force*, Prentice-Hall, 1960.

[HU 94] HU J., VAKILI P., YU G., "Optimality of hedging point policies in the production control of failure prone manufacturing systems", *Automatic Control, IEEE Transactions*, vol. 39, no. 9, pp. 1875–1880, 1994.

[IRA 02] IRAVANI, S.M.R., DUENYAS I., "Integrated maintenance and production control of a deteriorating production system", *IIE Transactions*, vol. 34, pp. 423–435, 2002.

[JAT 06] JATURONNATEE J., MURTHY D.N.P., BOONDISKULCHOK R., "Optimal preventive maintenance of leased equipment with corrective minimal repairs", *European Journal of Operational Research*, vol. 174, pp. 201–215, 2006.

[KEC 95] KECECIOGLU D., *Maintenability, Availability and Operational Readiness Engineering*, Prentice Hall, 1995.

[KEN 01] KENNE J.P., GHARBI A., "A simulation optimization approach in production planning of failure prone manufacturing systems", *Journal of Intelligent Manufacturing*, vol. 12, nos. 5–6, pp. 421–431, 2001.

[KIM 04] KIM C.S., DJAMALUDIN I., MURTHY D.N.P., "Warranty and discrete preventive maintenance", *Reliability Engineering and System Safety*, vol. 84, pp. 301–309, 2004.

[LAG 98] LAGARIAS J.C., REEDS J.A., WRIGHT M.H. *et al.*, "Convergence properties of the nelder-mead simplex method in low dimensions", *SIAM SIAM Journal on Optimization*, Vol. 9, pp. 112–147, 1998.

[LEE 85] LEE H.L., ROSENBLATT M.J., "Optimal inspection and ordering policies for Products with imperfect Quality", *IIE Transactions*, pp. 284–289, September 1985.

[LEE 05] LEE H.H., "A cost/benefit model for investments in inventory and preventive maintenance in an imperfect production system", *Computers & Industrial Engineering*, vol. 48, pp. 55–68, 2005.

[MUR 99] MURTHY, D.N.P., ASGHARIZADEH E., "Optimal decision making in a maintenance service operation", *European Journal of Operational Research*, vol. 116, pp. 259–273, 1999.

[NEL 65] NELDER JOHN A., MEAD R., "A simplex method for function minimization", *Computer Journal*, vol. 7, pp. 308–313, 1965.

[PIE 76] PIERSKALLA W.P., VOELKER J.A., "A survey of maintenance models: the control and surveillance of deteriorating systems", *Naval Research Logistics Quarterly*, vol. 23, pp. 355–388, 1976.

[QUA 02] OUALI M.S., REZG N., XIE X., "Maintenance préventive et optimisation des flux d'un système de production", *Journal Européen des Systèmes Automatisés*, vol. 36, no. 1, pp. 97–116, 2002.

[REZ 04] REZG N., XIE X., MATI Y., "Joint optimisation of preventive maintenance and inventory control in a production line using simulation", *International Journal of Production Research*, vol. 44, pp 2029–2046, 2004.

[RIC 05] RICHARD Y.K., FUNG T.C., "Multiagent supply chain planning and coordination architecture", *International Journal of Advanced Manufacturing Technology*, vol. 25, no. 7, pp. 811–819, 2005.

[ROS 86] ROSENBLATT M.J., LEE H.L., "Economic production Cycles with Imperfect Production Processes", *IEE Transactions*, pp. 48–55, March 1986.

[SAD 91] SADO G., SADO M.C., *Les plans d'expérience*, Afnor Technique 1991.

[SET 94] SETHI S.P., ZHOU X.Y., "Stochastic dynarnic job shops and hierarchical production planning", *IEEE Transactions on Automatic Control*, vol. 39, no. 10, pp. 2061–2076, 1994.

[SHA 11] SHAOMIN W., PHIL L., "Optimising age-replacement and extended non-renewing warranty policies in lifecycle costing", *International Journal of Economics*, vol. 130, pp. 262–267, 2011.

[SHE 81] SHERIF Y.S., SMITH M.L., "Optimal maintenance models for systems subject to failure- A review", *Naval Research Quaterly*, vol. 28, pp. 47–74, 1981.

[SHE 95] SHEU S.H., "Extented block replacement policy of a system subject to shocks", *IEEE Transactions on Reliability,* vol. 46, 1995.

[SIL 04] SILVA FILHO O.S., CEZARINO W., "An optimal production policy applied to a flow-shop manufacturing system", *Brazilian Journal of Operations and Production Management*, vol. 1, no. 1, pp. 73–92, 2004.

[SIL 05] SILVA FILHO O.S., "A stochastic constrained production planning problem with imperfect information of inventory", *Proceeding of IFAC World Congress,* Elsevier, 2005.

[SIN 96] SINGHAL J., SINGHAL K., "Alternate approaches to solving the Holt *et al.* model and to performing sensitivity analysis", *European Journal of Operational Research,* vol. 91, pp. 89–98, 1996.

[TEM 01] TEMPELMEIER H., "Performance evaluation of unbalanced flow lines with general distributed processing times, failures and imperfect production", *IIE Transactions*, vol. 33, no. 4, pp. 293–302, 2001.

[TUR 12] TURKI S., HAJEJ Z., REZG N., "Impact of delivery time on optimal production/delivery/maintenance planning", *IEEE International Conference on Automation Science and Engineering (CASE)*, Seoul, pp. 335–340, 2012.

[TUR 12a] TURKI S., HENNEQUIN S., SAUER N., "Perturbation analysis based-optimization for discrete flow model: a failure-prone manufacturing system with constant delivery time and stochastic demand", *Int. J. Advanced Operations Management,* vol. 4, nos. 1–2, pp. 124–153, 2012.

[TUR 12b] TURKI S., HAJEJ Z., REZG N., "Impact of delivery time on optimal production/delivery/maintenance planning", *IEEE International Conference on Automation Science and Engineering* (CASE), Seoul, pp. 335–340, 2012.

[TUR 13] TURKI S., HENNEQUIN S., SAUER N., "Perturbation analysis for continuous and discrete flow models: a study of the delivery time impact on the optimal buffer level", *International Journal of Production Research*, vol. 51, no. 13, pp. 4011–4044, 2013.

[VAL 89] VALDEZ-FLORES C., FELDMAN R., "A survey of preventive maintenance models for stochastically deterioring single-unit systems", *Naval Research Quaterly*, vol. 33, pp. 419–446, 1989.

[VAN 93] VAN RYZIN G., LOU S.X.C., GERSCHWIN B.S., "Production control for a two-machine system", *IIE Transactions*, vol. 25, no, 5, pp. 5–20, 1993.

[VAN 95a] VAN DER DUYN SCHOUTEN F.A., VANNESTE S.G., "Maintenance optimization of a production system with buffer capacity", *European Journal of Operational Research*, vol. 82, pp. 323–338, 1995.

[VAN 95b] VAN BRACHT E., "Performance analysis of a serial production line with machine breakdowns", *IEEE Symposium on Emerging Technologies & Factory Automation*, vol. 3, pp. 417–424, 1995.

[WAN 02] WANG H., "A survey of maintenance policies of deteriorating systems", *European Journal of Operational Research*, vol. 39, pp. 469–489, 2002.

[WON 08] WON, Y.Y., MURTHY D.N.P., JACK N., "Warranty servicing with imperfect repair", *International Journal of Production Economics*, vol. 111, pp. 159–169, 2008.

[XIE 93] XIE X.L., "Performance analysis of a transfer line with unreliable machines and finite buffers", *IIE Transactions,* vol. 25, no. 1, pp. 99–108, 1993.

[YAN 12] YANG B., LIANG G., TERO S. *et al.*, "Parameter-optimized simulated annealing for application mapping on networks-on-chip", *LION, volume 7219 of Lecture Notes in Computer Science,* Springer, pp. 307–322, 2012.

[YUH 07] YU-HUNG C., JIH-AN C., "Optimal age-replacement policy for repairable products under renewing free-replacement warranty", *International Journal of Systems Science*, vol. 38, no. 9, pp. 759–769, September 2007.

[YEH 07] YEH R.H., CHANG W.L., "Optimal threshold value of failure-rate for leased products with preventive maintenance actions", *Mathematical and Computer Modeling*, vol. 46, pp. 730–737, 2007.

Index

E, F, L, M

P, R

S, T, W

Other titles from

ISTE

in

Systems and Industrial Engineering – Robotics

2016

ANDRÉ Michel, SAMARAS Zissis
Energy and Environment
(Research for Innovative Transports Set - Volume 1)

AUBRY Jean-François, BRINZEI Nicolae, MAZOUNI Mohammed-Habib
Systems Dependability Assessment: Benefits of Petri Net Models (Systems Dependability Assessment Set - Volume 1)

BLANQUART Corinne, CLAUSEN Uwe, JACOB Bernard
Towards Innovative Freight and Logistics (Research for Innovative Transports Set - Volume 2)

COHEN Simon, YANNIS George
Traffic Management (Research for Innovative Transports Set - Volume 3)

MARÉ Jean-Charles
Aerospace Actuators 1: Needs, Reliability and Hydraulic Power Solutions

TORRENTI Jean-Michel, LA TORRE Francesca
Materials and Infrastructures 1 (Research for Innovative Transports Set - Volume 5A)
Materials and Infrastructures 2 (Research for Innovative Transports Set - Volume 5B)

WEBER Philippe, SIMON Christophe
Benefits of Bayesian Network Models
(Systems Dependability Assessment Set – Volume 2)

YANNIS George, COHEN Simon
Traffic Safety (Research for Innovative Transports Set - Volume 4)

2015

AUBRY Jean-François, BRINZEI Nicolae
Systems Dependability Assessment: Modeling with Graphs and Finite State Automata

BOULANGER Jean-Louis
CENELEC 50128 and IEC 62279 Standards

BRIFFAUT Jean-Pierre
E-Enabled Operations Management

MISSIKOFF Michele, CANDUCCI Massimo, MAIDEN Neil
Enterprise Innovation

2014

CHETTO Maryline
Real-time Systems Scheduling
Volume 1 – Fundamentals
Volume 2 – Focuses

DAVIM J. Paulo
Machinability of Advanced Materials

ESTAMPE Dominique
Supply Chain Performance and Evaluation Models

FAVRE Bernard
Introduction to Sustainable Transports

GAUTHIER Michaël, ANDREFF Nicolas, DOMBRE Etienne
Intracorporeal Robotics: From Milliscale to Nanoscale

MICOUIN Patrice
Model Based Systems Engineering: Fundamentals and Methods

MILLOT Patrick
Designing Human–Machine Cooperation Systems

NI Zhenjiang, PACORET Céline, BENOSMAN Ryad, REGNIER Stéphane
Haptic Feedback Teleoperation of Optical Tweezers

OUSTALOUP Alain
Diversity and Non-integer Differentiation for System Dynamics

REZG Nidhal, DELLAGI Sofien, KHATAD Abdelhakim
Joint Optimization of Maintenance and Production Policies

STEFANOIU Dan, BORNE Pierre, POPESCU Dumitru, FILIP Florin Gh.,
EL KAMEL Abdelkader
Optimization in Engineering Sciences: Metaheuristics, Stochastic Methods andDecision Support

2013

ALAZARD Daniel
Reverse Engineering in Control Design

ARIOUI Hichem, NEHAOUA Lamri
Driving Simulation

CHADLI Mohammed, COPPIER Hervé
Command-control for Real-time Systems

DAAFOUZ Jamal, TARBOURIECH Sophie, SIGALOTTI Mario
Hybrid Systems with Constraints

FEYEL Philippe
Loop-shaping Robust Control

FLAUS Jean-Marie
Risk Analysis: Socio-technical and Industrial Systems

FRIBOURG Laurent, SOULAT Romain
Control of Switching Systems by Invariance Analysis: Application to Power Electronics

GROSSARD Mathieu, REGNIER Stéphane, CHAILLET Nicolas
Flexible Robotics: Applications to Multiscale Manipulations

GRUNN Emmanuel, PHAM Anh Tuan
Modeling of Complex Systems: Application to Aeronautical Dynamics

HABIB Maki K., DAVIM J. Paulo
Interdisciplinary Mechatronics: Engineering Science and Research Development

HAMMADI Slim, KSOURI Mekki
Multimodal Transport Systems

JARBOUI Bassem, SIARRY Patrick, TEGHEM Jacques
Metaheuristics for Production Scheduling

KIRILLOV Oleg N., PELINOVSKY Dmitry E.
Nonlinear Physical Systems

LE Vu Tuan Hieu, STOICA Cristina, ALAMO Teodoro,
CAMACHO Eduardo F., DUMUR Didier
Zonotopes: From Guaranteed State-estimation to Control

MACHADO Carolina, DAVIM J. Paulo
Management and Engineering Innovation

MORANA Joëlle
Sustainable Supply Chain Management

SANDOU Guillaume
Metaheuristic Optimization for the Design of Automatic Control Laws

STOICAN Florin, OLARU Sorin
Set-theoretic Fault Detection in Multisensor Systems

2012

AÏT-KADI Daoud, CHOUINARD Marc, MARCOTTE Suzanne, RIOPEL Diane
Sustainable Reverse Logistics Network: Engineering and Management

BORNE Pierre, POPESCU Dumitru, FILIP Florin G., STEFANOIU Dan
Optimization in Engineering Sciences: Exact Methods

CHADLI Mohammed, BORNE Pierre
Multiple Models Approach in Automation: Takagi-Sugeno Fuzzy Systems

DAVIM J.Paulo
Lasers in Manufacturing

DECLERCK Philippe
Discrete Event Systems in Dioid Algebra and Conventional Algebra

DOUMIATI Moustapha, CHARARA Ali, VICTORINO Alessandro, LECHNER Daniel
Vehicle Dynamics Estimation using Kalman Filtering: Experimental Validation

GUERRERO José A, LOZANO Rogelio
Flight Formation Control

HAMMADI Slim, KSOURI Mekki
Advanced Mobility and Transport Engineering

MAILLARD Pierre
Competitive Quality Strategies

MATTA Nada, VANDENBOOMGAERDE Yves, ARLAT Jean
Supervision and Safety of Complex Systems

POLER Raul *et al.*
Intelligent Non-hierarchical Manufacturing Networks

TROCCAZ Jocelyne
Medical Robotics

YALAOUI Alice, CHEHADE Hicham, YALAOUI Farouk, AMODEO Lionel
Optimization of Logistics

ZELM Martin *et al.*
Enterprise Interoperability –I-EASA12 Proceedings

2011

CANTOT Pascal, LUZEAUX Dominique
Simulation and Modeling of Systems of Systems

DAVIM J. Paulo
Mechatronics

DAVIM J. Paulo
Wood Machining

GROUS Ammar
Applied Metrology for Manufacturing Engineering

KOLSKI Christophe
Human–Computer Interactions in Transport

LUZEAUX Dominique, RUAULT Jean-René, WIPPLER Jean-Luc
Complex Systems and Systems of Systems Engineering

ZELM Martin, *et al.*
Enterprise Interoperability: IWEI2011 Proceedings

2010

BOTTA-GENOULAZ Valérie, CAMPAGNE Jean-Pierre, LLERENA Daniel, PELLEGRIN Claude
Supply Chain Performance / Collaboration, Alignement and Coordination

BOURLÈS Henri, GODFREY K.C. Kwan
Linear Systems

BOURRIÈRES Jean-Paul
Proceedings of CEISIE'09

CHAILLET Nicolas, REGNIER Stéphane
Microrobotics for Micromanipulation

DAVIM J. Paulo
Sustainable Manufacturing

GIORDANO Max, MATHIEU Luc, VILLENEUVE François
Product Life-Cycle Management / Geometric Variations

LOZANO Rogelio
Unmanned Aerial Vehicles / Embedded Control

LUZEAUX Dominique, RUAULT Jean-René
Systems of Systems

VILLENEUVE François, MATHIEU Luc
Geometric Tolerancing of Products

2009

DIAZ Michel
Petri Nets / Fundamental Models, Verification and Applications

OZEL Tugrul, DAVIM J. Paulo
Intelligent Machining

PITRAT Jacques
Artificial Beings

2008

ARTIGUES Christian, DEMASSEY Sophie, NERON Emmanuel
Resources–Constrained Project Scheduling

BILLAUT Jean-Charles, MOUKRIM Aziz, SANLAVILLE Eric
Flexibility and Robustness in Scheduling

DOCHAIN Denis
Bioprocess Control

LOPEZ Pierre, ROUBELLAT François
Production Scheduling

THIERRY Caroline, THOMAS André, BEL Gérard
Supply Chain Simulation and Management

2007

DE LARMINAT Philippe
Analysis and Control of Linear Systems

DOMBRE Etienne, KHALIL Wisama
Robot Manipulators

LAMNABHI Françoise *et al.*
Taming Heterogeneity and Complexity of Embedded Control

LIMNIOS Nikolaos
Fault Trees

2006

FRENCH COLLEGE OF METROLOGY
Metrology in Industry

NAJIM Kaddour
Control of Continuous Linear Systems

Printed and bound by CPI Group (UK) Ltd, Croydon, CR0 4YY
15/09/2026
14962342-0001